PHYSICS RESEARCH AND TECHNOLOGY

AN ESSENTIAL GUIDE TO MAXWELL'S EQUATIONS

Physics Research and Technology

Additional books and e-books in this series can be found on Nova's website under the Series tab.

PHYSICS RESEARCH AND TECHNOLOGY

AN ESSENTIAL GUIDE TO MAXWELL'S EQUATIONS

CASEY ERICKSON
EDITOR

nova
science publishers
New York

Library of Congress Cataloging-in-Publication Data

ISBN: 978-1-53616-680-4

Published by Nova Science Publishers, Inc. † New York

CONTENTS

PREFACE

An Essential Guide to Maxwell's Equations first reviews the Ranada field line solutions of Maxwell's equations in a vacuum, describing a topologically non-trivial electromagnetic field, as well as their relation with the knot theory. Also, the authors present a generalization of these solutions to the non-linear electrodynamics recently published in the literature.

Next, this compilation reviews the gravitating electromagnetic field in the 1+3 formalism on a general hyperbolic space-time manifold, discussing the recent results regarding the existence of local field line solutions to the Einstein-Maxwell equations.

Lastly, the authors consider the existence of a weak solution to a class of an evolutionary Maxwell-Stokes type problem containing a p-curlcurl system in a multi-connected domain.

In Chapter 1, the authors review the Rañada field line solutions of Maxwell's equations in the vacuum, which describe a topologically non-trivial electromagnetic field, as well as their relation with the knot theory. Also, they present a generalization of these solutions to the non-linear electrodynamics recently published in the literature.

In Chapter 2, the authors are going to review the gravitating electromagnetic field in the 1+3 formalism on a general hyperbolic space-time manifold. The authors also discuss the recent results on the existence of the local field line solutions of the Einstein-Maxwell equations that generalize the Rañada solutions from the flat space-time. The global field line solutions do not always exist since the space-time manifold could impose obstructions to the global extension of various geometric objects necessary to build the fields. One example of a gravitating field line solution is the Kopiński-Natário field which is discussed in some detail.

Chapter 3 considers the existence of a weak solution to a class of an evolutionary Maxwell-Stokes type problem containing a p-curlcurl system in a multi-connected domain. Moreover, the authors show that the solution converges to a solution of the stationary Maxwell-Stokes type problem as the time tending to the infinity.

In: An Essential Guide to Maxwells Equations ISBN: 978-1-53616-680-4
Editor: Casey Erickson

Chapter 1

KNOTS AND THE MAXWELL EQUATIONS

Ion V. Vancea*
Grupo de Física Teórica e Matemática Física
Departamento de Física
Universidade Federal Rural do Rio de Janeiro
Seropédica, Rio de Janeiro, Brazil

Abstract

In this chapter, we review the Rañada field line solutions of Maxwell's equations in the vacuum, which describe a topologically non-trivial electromagnetic field, as well as their relation with the knot theory. Also, we present a generalization of these solutions to the non-linear electrodynamics recently published in the literature.

Keywords: Maxwell's equations, on-linear electrodynamics, Rañada solutions, knot solutions

1. INTRODUCTION

The discovery of the knot solutions of Maxwell's equations in the vacuum represents one of the most exciting results obtained recently in the modern classical electrodynamics. Since their first appearance in the seminal papers [1, 2, 3, 4],

*Corresponding Author's E-mail: ionvancea@ufrrj.br.

many interesting properties, applications, and generalizations of the knot electromagnetic fields have been discovered.

The knot solutions of Maxwell's equations can be described in terms of electric and magnetic field lines, and their topology can be given in terms of a pair of complex scalar fields that are interpreted as Hopf maps $S^3 \to S^2$ on the compactified space-like directions of the Minkowski space-time [4]-[6]. Among the properties studied up to now, one can cite: the relationship between the linked and the knotted electromagnetic fields discussed in [7, 8], the dynamics of the electric charges in topologically non-trivial electromagnetic backgrounds investigated in [10]-[13], and the topological quantization presented in [14]-[17]. In the last decade, the Rañada solutions were generalized in two ways. The first generalization is from knot fields to torus fields and was given in the works of Arrayas and Hoyos [18]-[20]. The second generalization is from the electromagnetic fields in vacuum to electromagnetic fields in matter. Several authors have showed that topological fields can be found in various areas of physics such as: fluid physics [22, 23], atmospheric physics [24], liquid crystals [25], plasma physics [26], optics [27, 28], and superconductivity [29]. (For a recent review of the knot solutions and their applications see [30] and the references therein). More recently, new and important generalizations of the topological electromagnetic fields have been made to the non-linear electrodynamics, the fluid physics [21]-[23] as well as to the gravitational physics [31] - [34]. Due to their wide range of applications in both physics and mathematics, the topological electromagnetic fields represent an important field of science and an active line of research.

In the present chapter, we are going to revisit the construction of the knot solutions of Maxwell's equations in Ranãda's approach. Also, we are going to briefly survey the different mathematical formulations of Maxwell's equations and to present the construction of the Hopf maps in electrodynamics. Of utmost importance for the understanding of the knot solutions is the factorization method of the 2-forms introduced by Bateman. We will review the application of Bateman's method to the topological electromagnetic fields. Finally, we are going to present the argument from [22, 23] where it was showed that knot solutions can also be found in the non-linear Born-Infeld electrodynamics and other non-linear generalizations of Maxwell's electrodynamics. We will focus our presentation on the electromagnetic fields in flat space-time. The generalization of the field line solutions to the gravitating electromagnetic fields is discussed in a different chapter of this volume [35]. The results reviewed here can be found in the original papers cited in the text. Other reviews are available too, most notably [30] to which we refer for an updated list of references. In the Appendix, we collect some basic mathematical properties of the Hopf mapping. We adopt throughout this chapter the natural units in which $c = 1$.

2. Maxwell's Equations

In this section, we will briefly review the formulation of Maxwell's electrodynamics in terms of differential forms in the three-dimensions Euclidian space and in the four-dimensional space-time (the covariant formulation). This is a well-known material which can be found in standard textbooks on classical electrodynamics such as [36, 37].

2.1. Maxwell's Equations in the Three-Dimensional Formulation

The two formulations of Maxwell's equations that are our concern in this paper are: the formulation in terms of differential forms and the covariant formulation. Both formulations are equivalent to the one in terms of three-dimensional vectors that is known from the undergraduate textbooks on electrodynamics, see e.g., [36].

Consider the following standard form of Maxwell's equations in the Heaviside units

$$\nabla \cdot \mathbf{D} = \rho\,, \tag{1}$$

$$\nabla \cdot \mathbf{B} = 0\,, \tag{2}$$

$$\nabla \times \mathbf{E} = -\frac{\partial \mathbf{B}}{\partial t}\,, \tag{3}$$

$$\nabla \times \mathbf{H} = \mathbf{J} + \frac{\partial \mathbf{D}}{\partial t}\,. \tag{4}$$

Here, $\mathbf{E}$ and $\mathbf{H}$ are the electric and magnetic intensities and $\mathbf{D}$ and $\mathbf{B}$ are the electric and magnetic flux densities, respectively. The sources of the electric and magnetic fields are the density of charge ρ and the density of vector current $\mathbf{J}$. All vectors are three-dimensional and they are defined on the entire $\mathbb{R}^3$ at each instant of time if no other conditions are imposed on the system.

The set of equations (1) - (4) can be written in terms of differential forms from either $\Omega^k(\mathbb{R}^3)$ or $\Omega^k(\mathbb{R}^{1,3})$ which denote the differential forms of rank k defined on the corresponding spaces. This abstract mathematical formulation is advantageous for at least two reasons - first, it provides a more economical formulation of the basic relations of the electromagnetism. Second, it provides the mathematical framework that is necessary for the generalization of the classical electromagnetism to curved space-times (see e.g., [37]). The differential forms associated with the three-dimensional electromagnetic field are given in the Table 1.

Table 1. Field content of Maxwell's equations

Vector notation	Form notation	Form rank	Field
E	$\mathcal{E}$	1 - form	electric intensity
H	$\mathcal{H}$	1 - form	magnetic intensity
D	$\mathcal{D}$	2 - form	electric flux density
B	$\mathcal{B}$	2 - form	magnetic flux density
ρ	$\mathcal{Q}$	3 - form	charge density
J	$\mathcal{J}$	2 - form	current density

It is easy to show that the differential forms from the Table 1 can be projected onto the orthonormal Cartesian basis of $\mathbb{R}^3$. The corresponding components are given by the following relations

$$\begin{aligned}
\mathcal{E} &= E_x dx + E_y dy + E_z dz \,, \\
\mathcal{H} &= H_x dx + H_y dy + H_z dz \,, \\
\mathcal{D} &= D_x dy \wedge dz + D_y dz \wedge dx + D_z dx \wedge dy \,, \\
\mathcal{B} &= B_x dy \wedge dz + B_y dz \wedge dx + B_z dx \wedge dy \,, \\
\mathcal{Q} &= \rho \, dx \wedge dy \wedge dz \,, \\
\mathcal{J} &= J_x dy \wedge dz + J_y dz \wedge dx + J_z dx \wedge dy \,.
\end{aligned} \tag{5}$$

The basis (dx, dy, dz) from the space $\Omega^1(\mathbb{R}^3)$ of 1-forms is associated by the canonical procedure to the basis (x, y, z) from $\mathbb{R}^3$. The second canonical basis $(dx \wedge dy, dy \wedge dz, dz \wedge dx)$ in the space $\Omega^2(\mathbb{R}^3)$ of 2-forms is constructed from the 1-forms (dx, dy, dz) by taking their wedge product which is defined as follows

$$\wedge : \Omega^k \times \Omega^s \to \Omega^{k+s} \,, \qquad (\omega, \sigma) \to \omega \wedge \sigma \,. \tag{6}$$

In the equation above, we have not specified which is the base space because the wedge product is defined in the same way on either $\mathbb{R}^3$ or $\mathbb{R}^{1,3}$.

In order to write Maxwell's equations in terms of differential forms, one has to recall that the exterior derivative d is defined as being the $\mathbb{R}$-linear map from $\Omega^k \to \Omega^{k+1}$ with the following properties

$$df = \partial_i f \, dx^i \,, \tag{7}$$

$$d^2\omega = 0 \,, \tag{8}$$

$$d(\omega \wedge \sigma) = d(\omega) \wedge \sigma + (-)^k \omega \wedge d(\sigma) \,. \tag{9}$$

Here, f is an arbitrary smooth function and ω and σ are arbitrary differential forms. Note that the exterior derivative is nilpotent, i. e. $d^2 = 0$.

The Hodge dual operation $\star$ is defined as the map from $\Omega^{n-k} \to \Omega^k$, where n is the dimension of the base manifold, that satisfies the following relation

$$\omega \wedge (\star\sigma) = \langle \omega, \sigma \rangle \mathrm{n} \,. \tag{10}$$

Here, n is an unitary vector and $\langle \cdot, \cdot \rangle$ is the scalar product. In particular, the action of the Hodge star operator on the components of an arbitrary k-form is given by the following equations [37]

$$\omega = \frac{1}{k!} \omega_{i_1,\ldots,i_k} dx^{i_1} \wedge \cdots \wedge dx^{i_k} = \sum_{i_1 < \cdots < i_k} \omega_{i_1,\ldots,i_k} dx^{i_1} \wedge \cdots \wedge dx^{i_k} \,, \tag{11}$$

$$(\star\,\omega) = \frac{1}{(n-k)!} (\star\,\omega)_{i_{k+1},\ldots,i_n} dx^{i_{k+1}} \wedge \cdots \wedge dx^{i_n} \,. \tag{12}$$

The above equations can be easily particularized to the three-dimensional case. For example, the Hodge duals to the elements of the canonical basis in the space of 1-forms are given by the following relations

$$\star dx = dy\, dz \,, \qquad \star dy = dz\, dx \,, \qquad \star dz = dx\, dy \,. \tag{13}$$

Similarly, the Hodge duals of the elements of the canonical basis in the space of 1-forms are obtained from the equation (12) and they are given by the following relations

$$\star\, dy\, dz = dx \,, \qquad \star\, dz\, dx = dy \,, \qquad \star\, dx\, dy = dz \,. \tag{14}$$

The equations (13) and (14) illustrate the more general property of involution $\star(\star\omega) = \omega$ in the three-dimensional Euclidean space. By using the equations (12), (13) and (14), one can show that the exterior derivative d can be decomposed in the canonical basis as follows

$$d = (\partial_x dx + \partial_y dy + \partial_z dz) \wedge \, . \tag{15}$$

By using the exterior derivative from the equation (15), one can write the Maxwell equations in terms of the three-dimensional differential forms. The result is given by the following set of equations

$$d\mathcal{D} = \mathcal{Q} \, , \tag{16}$$
$$d\mathcal{B} = 0 \, , \tag{17}$$
$$d\mathcal{E} = -\partial_t \mathcal{B} \, , \tag{18}$$
$$d\mathcal{H} = \mathcal{J} + \partial_t \mathcal{D} \, . \tag{19}$$

It is easy to recognize in the equations (16) - (19) the familiar laws of the electromagnetism. They allow one to write the electromagnetic field in terms of potentials Φ and $\mathcal{A}$. That is made possible by the *Poincar's theorem*, that states that on a contractible manifold all closed forms ($d\omega = 0$) are exact, i. e. for any exact form ω there exists a form σ such that $\omega = d\sigma$ [37]. By applying this mathematical result, one can show that the electric and magnetic 1-forms can be written as follows

$$\mathcal{E} = -d\Phi - \partial_t \mathcal{A} \, , \tag{20}$$
$$\mathcal{B} = d\mathcal{A} \, . \tag{21}$$

It is an simple exercise to show that the equations (16) - (19) are invariant under the following gauge transformations

$$\Phi \rightarrow \Phi' = \Phi - \partial_t \Lambda \, , \tag{22}$$
$$\mathcal{A} \rightarrow \mathcal{A}' = \mathcal{A} + d\Lambda \, , \tag{23}$$

where Λ is an arbitrary smooth function that plays the role of the gauge parameter.

The equations (16) - (19) describe the dynamics of the electromagnetic field alone. Their solutions are given in terms of charge and current densities, respectively, which are non-dynamical objects. As usual, the change of the source state is given by Newton's second law written for the Lorentz force, which has the following form

$$\mathcal{F}_L = \rho\mathcal{E} - \iota_{(\star\mathcal{J})}\mathcal{B}\,. \tag{24}$$

The interior product ι is defined as the contraction between a differential form from the space Ω^k and a vector field X. The interior product lowers the form degree by one, and its image belongs to the space Ω^{k-1}. For example, if $\omega \in \Omega^2$ is a 2-form, its interior product with the vector field X is the 1-form $\iota_X\omega$ for which the following equality holds

$$\iota_X\omega(Y) = \omega(X, Y)\,, \tag{25}$$

where Y is an arbitrary smooth vector field. The equation (25) is the last equation needed to represent the dynamics of the electromagnetic field and its sources in terms of differential forms. Since in the rest of this chapter we will investigate the topological solutions of the electromagnetic field in the vacuum, that is away from its sources, we will ignore the Lorentz force as well as the charge and current densities.

2.2. Maxwell Equations in Covariant Formulation

The geometric properties of the electromagnetic field are highlighted in the formulation of Maxwell's equations in terms of differential forms on $\mathbb{R}^3$. However, the fundamental symmetry of Maxwell's equations, which is the invariance under the Lorentz transformations, is explicitly displayed only in the Minkowski space-time $\mathbb{R}^{1,3}$. Therefore, it is important to write the Maxwell equations in terms of differential forms on $\mathbb{R}^{1,3}$. To this end, the canonical basis $\{dx^i\}$ on $\mathbb{R}^3$ must be extended to the corresponding basis on $\mathbb{R}^{1,3}$, denoted by $\{dx^\mu\} = \{dx^0 = dt, dx^i\}$, by including the time-like 1-form dx^0. Here, we are using the indices $\mu, \nu = 0, 1, 2, 3$ to denote the geometrical objects and their components in the Minkowski space-time. The electromagnetic 2-form field F is defined by the following relation

$$F = B + E \wedge dx^0\,. \tag{26}$$

In this notation, the components of F in the basis $\{dx^\mu \wedge dx^\nu\}$ are the same as the components of the electromagnetic rank-2 antisymmetric tensor $F_{\mu\nu}$, namely

$$F = \frac{1}{2}F_{\mu\nu}dx^\mu \wedge dx^\nu\,. \tag{27}$$

The source of the electromagnetic field can also be written in terms of differential forms in four dimensions and it is expressed by the 1-form $J = J_\mu dx^\mu$ whose components are equal to the projections of the four-current $J^\mu = (\rho, \mathbf{J})$ onto the space-time directions. By using these mathematical objects, one can write down the Maxwell equations in their most compact form as follows

$$dF = 0\,, \tag{28}$$

$$\star d \star F = J\,. \tag{29}$$

It is easy to show that the homogeneous equation (28) corresponds to the magnetic Gauss law and to the Faraday law combined into one equation, while the inhomogeneous equation (29) is the same as the electric Gauss law and the Maxwell-Ampére law packed in to a single mathematical relation.

The equations (28) and (29) fix to some extent the geometrical properties of the electromagnetic 2-form. Indeed, from the geometrical point of view, the field F is a closed 2-form due to the equation (28). From this property and from the nilpotency of the exterior derivative, i. e. $d^2 = 0$, we can derive the action of the differential form on the electromagnetic field which is given by the following equation

$$dF = dB + dE \wedge dx^0\,. \tag{30}$$

Recall that the exterior derivative on $\mathbb{R}^{1,3}$ can be decomposed in to two terms: the exterior derivative $\mathbf{d}$ on the spatial subspace $\mathbb{R}^3 \in \mathbb{R}^{1,3}$ and the 1-form dx^0 along the time-like direction $\mathbb{R}$. This decomposition is given by the following equation

$$d = dx^0 \wedge \partial_0 + \mathbf{d}\,. \tag{31}$$

By plugging the equations (31) and (30) into the equation (28), one obtains the following set of equations

$$\mathbf{d}E + \partial_0 B = 0\,, \tag{32}$$

$$\mathbf{d}B = 0\,. \tag{33}$$

The equations (32) and (33) are the homogeneous Maxwell equations written in terms of 1-forms on the Euclidean space $\mathbb{R}^3$.

The three-dimensional inhomogeneous equations can be derived from the four-dimensional system, too. The outline of the derivation is the following. Firstly, note that the electromagnetic Hodge dual form $\star F$ can be obtained from F by making the following replacements

$$E_j \to -B_j\,, \qquad B_j \to E_j\,. \tag{34}$$

Secondly, decompose the form $\star F$ as follows

$$\star F = \star E - \star B \wedge dx^0 \,, \tag{35}$$

where $\star$ denotes the three-dimensional Hodge star operator. Thirdly, calculate the sequence of operations from the equation (29). Then one obtains the following set of inhomogeneous equations

$$\star \mathbf{d} \star E = \rho \,, \tag{36}$$

$$\star \mathbf{d} \star B - \partial_0 E = \mathcal{J} \,. \tag{37}$$

This concludes the derivation of the three-dimensional Maxwell equations in terms of differential forms from the four dimensional formulation.

Let us make some observations about the electromagnetic fields away from their sources. In this case, the fields propagate in the vacuum. Their dynamics is still given by the equations (28) and (29), but with the supplementary condition $J = 0$. Due to this last relation, the symmetries of Maxwell's equations are enhanced to the group $SO(1,3) \times U(1) \times \mathcal{D}$, where $\mathcal{D}$ denotes the electromagnetic duality symmetry defined by the following transformations

$$F \leftrightarrow \star F \,. \tag{38}$$

The relation (38) implies that the 2-form F can be written as a sum between a self-dual form F_+ and an anti self-dual form F_-. The corresponding equations are

$$F = F_+ + F_- \,, \qquad \star F_\pm = \pm i F_\pm \,. \tag{39}$$

The last of the two equations from above is the result of the self-duality relation for 2-forms in the Minkwoski space-time: $\star \star \, \omega = -\omega$.

We end this section by observing that the formulation of the classical electrodynamics in terms of differential forms is completely equivalent with the formulation in terms of three-dimensional vector fields. However, each framework has its own advantages. The vector approach is useful for the visualization of the spatial distribution of fields and sources. The differential forms are interesting because they provide more economical equations, a deeper view of the symmetries and more information about the geometrical and topological properties of the electromagnetic fields.

3. KNOTS IN MAXWELL'S ELECTRODYNAMICS

In this section, we are going to review the field line solutions of Maxwell's equations and their relationship with the Hopf knots. These solutions were discovered by Rañada and were communicated in [2, 3]. An important study of the Hopf fibration in the context of the classical electromagnetism can be found in the early pioneering work by Trautman [1]. In our presentation, we follow the very good indepth review of these solutions and their applications given in the reference [30].

3.1. Ranãda Solutions

The Ranãda solutions form a particular class of *field line solutions* whose main feature is that the field lines completly characterize the electromagnetic field. In general, one can associate a field line to any smooth vector field through its integral flow. The path formed by the field line is tangent in each of its points to the vector obtained by taking the value of the vector field at that point. The picture of all field lines at a given instant of time can be used to visualize the state of the vector field at that instant.

As we can see from the equations (28) and (29), the properties of the sources determine the physical and geometrical properties of the electromagnetic field. In particular, the field lines of the electromagnetic field can have a very complex structure for non-trivial sources and can take a simple form in the vacuum.

In order to construct an electromagnetic field in terms of its field lines, we need a formal description of the latter. That can be given in terms of two smooth scalar complex fields on $\mathbb{R}^3$, namely

$$\phi : \mathbb{R}^3 \to \mathbb{C}\,, \qquad \theta : \mathbb{R}^3 \to \mathbb{C}\,. \tag{40}$$

By introducing the above functions, one can interpret the electric and the magnetic field lines as level lines of θ and ϕ. In the covariant formulation, the corresponding electromagnetic fields take the following form [30]

$$F_{\mu\nu} = g(\bar{\phi}, \phi)\left(\partial_\mu \bar{\phi}\, \partial_\nu \phi - \partial_\nu \bar{\phi}\, \partial_\mu \phi\right)\,, \tag{41}$$

$$\star F_{\mu\nu} = f(\bar{\theta}, \theta)\left(\partial_\mu \bar{\theta}\, \partial_\nu \theta - \partial_\nu \bar{\theta}\, \partial_\mu \theta\right)\,, \tag{42}$$

where g and f are smooth functions on θ and ϕ, and the dual electromagnetic field has the following form

$$\star F_{\mu\nu} = \frac{1}{2}\epsilon_{\mu\nu\rho\sigma} F^{\rho\sigma}\,. \tag{43}$$

The components of the electromagnetic tensor $F_{\mu\nu}$ contain the electric and magnetic vector fields **E** and **B**. These can be obtained from the following relations

$$F = \frac{1}{2} F_{\mu\nu} dx^\mu \wedge dx^\nu = -\varepsilon_{jkl} B_j dx^k \wedge dx^l + E_j dx^j \wedge dx^0 \,, \tag{44}$$

$$\star F = \frac{1}{2} \star F_{\mu\nu} dx^\mu \wedge dx^\nu = \varepsilon_{jkl} E_j dx^k \wedge dx^l + B_j dx^j \wedge dx^0 \,. \tag{45}$$

Note that the field line solutions (41) and (42) are given in the covariant formulation. Although they are explicitly Lorentz covariant and electromagnetically dual to each other, it is almost impossible to visualize the geometry of the electromagnetic field from the equations (41) and (42). This problem can be solved by invoking the equations (44) and (45) and by giving concrete values to the arbitrary functions g and f. The first solution with the knot topology obtained in this way is the Ranãda solution from [2, 3]. It has the following electric and magnetic decomposition

$$E_j = \frac{\sqrt{a}}{2\pi i} \left(1 + |\theta|^2\right)^{-2} \varepsilon_{jkl}\, \partial_k \bar{\theta}\, \partial_l \theta \,, \tag{46}$$

$$B_j = \frac{\sqrt{a}}{2\pi i} \left(1 + |\phi|^2\right)^{-2} \varepsilon_{jkl}\, \partial_k \bar{\phi}\, \partial_l \phi \,. \tag{47}$$

The electromagnetic duality imposes some constraints on the fields θ and ϕ. These constraints can be easily found by substituting the functions $g(\bar{\phi}, \phi)$ and $f(\bar{\theta}, \theta)$ from the equations (46) and (47) into the equation (38), which leads to the following equations [30]

$$\left(1 + |\phi|^2\right)^{-2} \varepsilon_{jmn} \partial_m \phi \partial_n \bar{\phi} = \left(1 + |\theta|^2\right)^{-2} \left(\partial_0 \bar{\theta} \partial_j \theta - \partial_0 \theta \partial_j \bar{\theta}\right) \,, \tag{48}$$

$$\left(1 + |\theta|^2\right)^{-2} \varepsilon_{jmn} \partial_m \bar{\theta} \partial_n \theta = \left(1 + |\phi|^2\right)^{-2} \left(\partial_0 \bar{\phi} \partial_j \phi - \partial_0 \phi \partial_j \bar{\phi}\right) \,. \tag{49}$$

The equations (48) and (49) form a set of independent non-linear partial differential equations. The electric and magnetic field lines are the level curves of the solutions of the equations (48) and (49).

As one can see from the equations (46) and (47), the Ranãda fields correspond to particular functions g and f. Recall that the *null field* solutions of Maxwell's equations in vacuum are defined by the following equations

$$E_j B_k \delta_{jk} = 0 \,, \tag{50}$$

$$\delta_{jk} \left(E^j E^k - B^j B^k\right) = 0 \,. \tag{51}$$

Then it is easy to verify that Ranãda fields $\mathbf{E}$ and $\mathbf{B}$ satisfy the first null field equation (50), but do not satisfy the second one. From that, we can conclude that the electric and magnetic fields are orthogonal to each other and so are their field lines.

The parametrization of the electromagnetic 2-forms F and $\star F$ in terms of ϕ and θ is not unique according to the equations (46) and (47). Other possibility is to use the so called *Clebsch parametrization*. The main mathematical tool to construct it is the Darboux theorem [37] that is satisfied by both F and $\star F$. By using the Darboux theorem, we can write F and $\star F$ in terms of four canonical 1-forms $d\sigma^a$, $d\xi^a$, $d\rho^a$ and $d\zeta^a$ [30] as follows

$$F = \delta_{ab}\, d\sigma^a \wedge d\xi^b\,, \tag{52}$$

$$\star F = \delta_{ab}\, d\rho^a \wedge d\zeta^b\,, \tag{53}$$

$$d\sigma^a \wedge d\xi^a = d\rho^a \wedge d\zeta^a = -i\tau_2\,, \tag{54}$$

where the indices $a, b = 1, 2$ enumerate the 1-forms and τ_2 is the Pauli matrix. The Clebsch representation of the electromagnetic field takes a simpler form for fields that satisfy the equation (50), namely

$$F = d\sigma \wedge d\xi\,, \tag{55}$$

$$\star F = d\rho \wedge d\zeta\,. \tag{56}$$

Another important parametrization of the electromagnetic field is given in terms of *Euler potentials* which are smooth scalar real fields on the Minkowski space-time [30]

$$\alpha_a : \mathbb{R}^3 \to \mathbb{C}\,, \qquad \beta_a : \mathbb{R}^3 \to \mathbb{C}\,, \tag{57}$$

where $a = 1, 2$. The electric and magnetic fields have the following form

$$E_j = \varepsilon_{jkl}\, \partial_k \beta_2\, \partial_l \beta_1\,, \tag{58}$$

$$B_j = \varepsilon_{jkl}\, \partial_k \alpha_2\, \partial_l \alpha_1\,. \tag{59}$$

By comparing the two sets of equations (46)-(47) and (58)-(59) with each other, one can easily find the following relation among the parameters of the Ranãda and Euler parametrizations, respectively,

$$\beta_1 = \left(1 + |\theta|^2\right)^{-1}, \qquad \beta_2 = \frac{1}{2\pi}\arg(\theta)\,, \tag{60}$$

$$\alpha_1 = \left(1 + |\phi|^2\right)^{-1}, \qquad \alpha_2 = \frac{1}{2\pi}\arg(\phi)\,. \tag{61}$$

These representations are useful to express different properties of the field line solutions. Other representations can be found for more general topological electromagnetic fields. But before discussing this case, let us take a closer look at the Ranãda fields.

3.2. Electromagnetic Knot Fields

As we have seen above, the electromagnetic field characterized by the equations (41) and (42) is parametrized by g and f. In particular, by choosing these functions as in the equations (46) and (47), the field line solutions display non-trivial topological properties. In order to see that, we note that ϕ and θ should be chosen such that the electromagnetic field and the observables constructed from it, e.g., energy, linear momentum, and angular momentum, be finite. Therefore, if the field is defined in the full $\mathbb{R}^3$, the regularity conditions to be imposed on the complex functions are the following

$$|\phi(x)| \to 0 \text{ and } |\theta(x)| \to 0 \text{ if } |\mathbf{x}| \to \infty \,. \tag{62}$$

The boundary conditions (62) imply that the defining domain of ϕ and θ is the compactification $\mathbb{R}^3 \cup \{\infty\} = S^3$ at any given value of t. Since ϕ and θ take values in the compactification $\mathbb{C} \cup \{\infty\} = S^2$, they can be viewed as two families of one-parameter maps $\phi(x) = \{\phi_t(\mathbf{x})\}_{t\in\mathbb{R}}$ and $\theta(x) = \{\theta_t(\mathbf{x})\}_{t\in\mathbb{R}}$ from S^3 to S^2. Here, it is convenient to use the vector notation to discuss separately the electric and the magnetic field lines, respectively.

The fields $\mathbf{E}$ and $\mathbf{B}$ can be written in terms of electromagnetic potentials $\mathbf{C}$ and $\mathbf{A}$ in a symmetric way

$$E_j = \varepsilon_{jkl}\partial_k C_l \,, \qquad B_j = \varepsilon_{jkl}\partial_k A_l \,. \tag{63}$$

Note that $\mathbf{C}$ is dependent on $\mathbf{A}$, as they are related by the following equation

$$\varepsilon_{jmn}\partial_m C_n = -\partial_0 A_j \,. \tag{64}$$

However, it is convenient to keep $\mathbf{C}$ explicit, as it makes the electric-magnetic duality more symmetric. The Chern-Simons integrals associated to the electric and magnetic fields and their potentials can be used to define the helicities of the electromagnetic field as follows

$$H_{ee} = \int d^3x\, \delta_{ij} E_i C_j = \int d^3x\, \varepsilon_{jkl} C_j \partial_k C_l\,, \tag{65}$$

$$H_{mm} = \int d^3x\, \delta_{ij} B_i A_j = \int d^3x\, \varepsilon_{jkl} A_j \partial_k A_l\,, \tag{66}$$

$$H_{em} = \int d^3x\, \delta_{ij} B_i C_j = \int d^3x\, \varepsilon_{jkl} C_j \partial_k A_l\,, \tag{67}$$

$$H_{mm} = \int d^3x\, \delta_{ij} E_i A_j = \int d^3x\, \varepsilon_{jkl} A_j \partial_k C_l\,. \tag{68}$$

The pure electric and magnetic helicites given above are the Hopf indexes of the corresponding electric and magnetic field lines (see the Appendix). If one substitutes the equations (63) into Maxwell's equations (1) - (4) in the vacuum, one obtains the following set of equations

$$\varepsilon_{jmn}\partial_m \left(\frac{\partial A_n}{\partial x^0} + \varepsilon_{nrs}\partial_r C_s \right) = 0\,, \tag{69}$$

$$\varepsilon_{jmn}\partial_m \left(\frac{\partial C_n}{\partial x^0} - \varepsilon_{nrs}\partial_r A_s \right) = 0\,. \tag{70}$$

From the vector calculus, we conclude that there are two scalar functions κ_1 and κ_2 such that

$$\frac{\partial A_j}{\partial x^0} + \varepsilon_{jmn}\partial_m C_n = \partial_j \kappa_1\,, \tag{71}$$

$$\frac{\partial C_j}{\partial x^0} - \varepsilon_{jmn}\partial_m A_n = \partial_j \kappa_2\,. \tag{72}$$

By using the equations (69) - (72), the following equations are obtained

$$\delta_{mn}\frac{\partial (A_m B_n)}{\partial x^0} + 2\delta_{mn} E_m B_n - \partial_k \left(\varepsilon_{krs} A_k E_s - \kappa_1 B_k \right) = 0\,. \tag{73}$$

$$\delta_{mn}\frac{\partial (C_m E_n)}{\partial x^0} + 2\delta_{mn} E_m B_n + \partial_k \left(\varepsilon_{krs} C_k B_s + \kappa_2 E_k \right) = 0\,. \tag{74}$$

From these, it follows that

$$\frac{\partial (H_{mm} - H_{ee})}{\partial x^0} + 4\int d^3x \delta_{mn} E_m B_n = 0\,, \tag{75}$$

$$\frac{\partial (H_{mm} + H_{ee})}{\partial x^0} = 0\,. \tag{76}$$

The equations (75) and (76) show that the pure electric and magnetic helicities are conserved if the fields satisfy the first null field condition (50).

Let us quote here the explicit Hopfion solution obtained in [6]. The fields ϕ and θ have the following form

$$\phi = \frac{(ax_1 - x_0 x_3) + i(ax_2 + x_0(a-1)}{(ax_3 + x_0 x_1) + i[a(a-1) - x_0 x_1]}, \tag{77}$$

$$\theta = \frac{[ax_2 + x_0(a-1)] + i(ax_3 + x_0 x_1)}{(ax_1 - x_0 x_3) + i[a(a-1) - x_0 x_2]}, \tag{78}$$

where x_μ's are dimensionless coordinates and

$$a = \frac{r^2 - x_0^2 + 1}{2}, \qquad r = \sqrt{\delta_{mn} x_m^2 x_n^2}\,. \tag{79}$$

The electric and magnetic fields derived from the scalars given by the equations (77) and (78) have the following form

$$\mathbf{E} = \frac{1}{\pi} \frac{q\mathbf{H}_1 - p\mathbf{H}_2}{(a^2 + x_0^2)^3}, \tag{80}$$

$$\mathbf{B} = \frac{1}{\pi} \frac{q\mathbf{H}_1 + p\mathbf{H}_2}{(a^2 + x_0^2)^3}, \tag{81}$$

where

$$\begin{aligned} \mathbf{H}_1 &= (x_2 + x_0 - x_1 x_3)\,\mathbf{e}_1 - [x_1 + (x_2 + x_0)\,x_3]\,\mathbf{e}_2 \\ &+ \frac{1}{2}\left[-1 - x_3^2 + x_1^2 + (x_2 + x_0)^2\right]\mathbf{e}_3 \end{aligned} \tag{82}$$

$$\begin{aligned} \mathbf{H}_2 &= +\frac{1}{2}\left[1 + x_1^2 - x_3^2 - (x_2 + x_0)^2\right]\mathbf{e}_1 + [-x_3 + (x_2 + x_0)\,x_1]\,\mathbf{e}_2 \\ &+ (x_2 + x_0 + x_1 x_3)\,\mathbf{e}_3\,, \end{aligned} \tag{83}$$

and

$$p = x_0\left(x_0^2 - 3a^2\right), \qquad q = a\left(a^2 - 3x_0^2\right). \tag{84}$$

Here, we have denoted by $\{\mathbf{e}_l\}$ the unit vectors of the Cartesian basis in the spatial directions. The Hopf indices of the above Hopfion solution are $H(\phi) = H(\theta) = 1$ [30].

Another example of Hopfion is given by the Hopf map defined by the following scalar fields

$$\phi = \frac{2(x_1 + ix_2)}{2x_3^2 + i(\delta_{mn}x^m x^n - 1)}, \tag{85}$$

$$\theta = \bar{\phi}. \tag{86}$$

The electromagnetic field corresponding to these scalars has the following components

$$E_m = \frac{1}{\sqrt{(2\pi)^3}} \int d^3k \left[P_m(k_j)\cos\left(\eta_{\mu\nu}k^\mu x^\nu\right) - Q_m(k_j)\sin\left(\eta_{\mu\nu}k^\mu x^\nu\right)\right], \tag{87}$$

$$B_m = \frac{1}{\sqrt{(2\pi)^3}} \int d^3k \left[P_m(k_j)\cos\left(\eta_{\mu\nu}k^\mu x^\nu\right) + Q_m(k_j)\sin\left(\eta_{\mu\nu}k^\mu x^\nu\right)\right], \tag{88}$$

where

$$\mathbf{P} = \frac{e^{-k_0}}{\sqrt{2\pi}}\left(-\frac{k_1k_3}{k_0}, \frac{k_0k_2 + k_2^2 + k_3^2}{k_0}, -\frac{k_0k_1 + k_1k_2}{k_0}\right), \tag{89}$$

$$\mathbf{Q} = \frac{e^{-k_0}}{\sqrt{2\pi}}\left(-\frac{k_0k_2 + k_1^2 + k_2^2}{k_0}, \frac{k_1k_3}{k_0}, \frac{k_0k_3 + k_2k_3}{k_0}\right). \tag{90}$$

The fields given by the equations (87) and (88) describe a particular wave packet that travels along the x_3 axis and has the following energy, linear momentum, and angular momentum densities

$$E = 2, \qquad \mathbf{p} = (0,0,1), \qquad \mathbf{L} = (0,0,1). \tag{91}$$

The exact values of E, $\mathbf{p}$ and $\mathbf{L}$ are calculated in a dimensionless parametrization of the space-time coordinates. From the equations (91), it follows that the mass of the wave packet is finite and it has the value $m^2 = 3$.

4. Electric and Magnetic Knots in Bateman Parametrization

In the previous section, we have discussed several parametrizations of the field line solutions. In this section, we present the factorized parametrization of the self-dual electromagnetic fields in terms of 1-forms given by Bateman in [39]. This representation is useful for the generalization of the field line solutions to the gravitating electromagnetic fields [32] as well as to the non-linear electromagnetism [22, 23]. For extensive reviews of the properties of the Hopfions in the Bateman representation see [20, 30, 40].

Let us recall the covariant form of Maxwell's equation in the vacuum given by the equations (28) and (29), namely

$$dF = 0\,, \tag{92}$$

$$d \star F = 0\,. \tag{93}$$

The equation (92) states that the electromagnetic 2-form field is closed. Then F can be written in terms of two scalar complex functions

$$\alpha : \mathbb{R}^3 \to \mathbb{C}\,, \qquad \beta : \mathbb{R}^3 \to \mathbb{C}\,, \tag{94}$$

as follows

$$F = d\alpha \wedge d\beta\,. \tag{95}$$

It is relevant to write the above equation in terms of the electric and magnetic fields. To this end, we use the components of the electromagnetic 2-form F which are the same as the components of the electromagnetic tensor $F_{\mu\nu}$. It follows from the equation (95) that $F_{\mu\nu}$ satisfies the following relation

$$i\varepsilon^{\mu\nu\rho\sigma}F_{\rho\sigma} - 2F^{\mu\nu} = 2\varepsilon^{\mu\nu\rho\sigma}\partial_\rho\alpha\partial_\sigma\beta\,. \tag{96}$$

By definition, the electric and magnetic fields are given by the following equations

$$-E^m = F^{0m}\,, \qquad B^m = \frac{1}{2}\varepsilon^{mnp}F_{np}\,. \tag{97}$$

Some algebraic manipulations of the equation (96) lead to the following relation between the complex scalars

$$B_m - iE_m = i\left(\partial_0\alpha\partial_m\beta - \partial_0\beta\partial_m\alpha\right)\,. \tag{98}$$

We know from the equation (39) that the electromagnetic field in the vacuum is either self-dual or anti self-dual according to the eigenvalues of the $\star$ operation. That implies that the functions α and β must obey the following constraint

$$\nabla\alpha \times \nabla\beta = \pm i\left(\partial_0\alpha\nabla\beta - \partial_0\beta\nabla\alpha\right)\,. \tag{99}$$

In order to understand the geometry of the electric and magnetic fields on $\mathbf{R}^3$, it is useful to introduce the Riemann-Silberstein vector

$$\mathbf{F} = \mathbf{B} \pm i\mathbf{E}\,, \tag{100}$$

where $\mathbf{E}$ and $\mathbf{B}$ can be complex. Also, it is required that $\mathbf{F}$ be a solution of Bateman's equation

$$\delta_{mn} F_m F_n = 0\,. \tag{101}$$

Note that in general the norm of the field $\mathbf{F}$ is non-zero

$$\delta_{mn} \bar{F}_m F_n \neq 0\,, \tag{102}$$

where the bar stands for the complex conjugate. One can easily write the equations (101) and (102) on components. The result is the following

$$\delta_{mn}\left(B_m B_n - E_m E_n\right) \pm 2i\delta_{mn} E_m B_n = 0\,, \tag{103}$$

$$\delta_{mn}\left(\bar{B}_m B_n + \bar{E}_m E_n\right) = 0\,. \tag{104}$$

If the vector fields $\mathbf{E}$ and $\mathbf{B}$ are real, the left-hand side of the equation (103) is an invariant of the electromagnetic field and the left-hand side of the equation (104) is the energy density of the electromagnetic field. The equations (103) and (104) define the so called *null fields* [19].

It is easy to verify that the Bateman solutions conserve the energy, momentum and the angular momentum of the electromagnetic field. These conservation laws follow from the Lorentz symmetry. Also, the $U(1)$ symmetry of Maxwell's equations implies that there is a conserved four-current whose components are given by the following relations

$$\rho = \frac{1}{2}\delta_{mn}\left(E_m E_n + B_m B_n\right)\,, \tag{105}$$

$$J_k = \varepsilon_{kmn} E_m B_n\,. \tag{106}$$

Beside the conserved quantities discussed above, there are new topologically conserved charges, the helicities from the equations (65)-(66). The Bateman equations admit solution with knot as well as toric topologies. In the Bateman parametrization, different solutions can be related to each other, or derived from each other, due to the following important theorem [28]:

Let α and β be two smooth complex scalar fields on M that satisfy the Bateman relation (99). Then for any two arbitrary smooth complex functions f and g defined on $\mathbb{C}^2$, the following 2-form exists

$$\mathcal{F} := df(\alpha, \beta) \wedge dg(\alpha, \beta)\,, \tag{107}$$

and $\mathcal{F}$ has the following properties:

$$d\mathcal{F} = 0\,, \tag{108}$$

$$\star\mathcal{F} = \pm i\mathcal{F}\,. \tag{109}$$

There are several known solutions in the literature constructed with the Bateman method. Let us cite here the Hopfion obtained in [19] whose complex functions are given by the following relations

$$\alpha = \frac{-x_0^2 + \delta_{mn}x^m x^n - 1 + 2ix_3}{-x_0^2 + \delta_{mn}x^m x^n + 1 + 2ix_0}\,, \tag{110}$$

$$\beta = \frac{x_1 - ix_2}{-x_0^2 + \delta_{mn}x^m x^n + 1 + 2ix_0}\,. \tag{111}$$

The functions α and β given above satisfy the following relation

$$|\alpha|^2 + |\beta|^2 = 1\,. \tag{112}$$

The electromagnetic vector $\mathbf{F}$ that is obtained from α and β has the following form

$$\mathbf{F} = \frac{4}{(-x_0^2 + \delta_{mn}x^m x^n + 1 + 2ix_0)^3} \times \begin{bmatrix} (x_0 - x_1 - x_3 + i\,(x_2 - 1))\,(x_0 + x_1 - x_3 - i\,(x_2 + 1)) \\ -i\,(x_0 - x_2 - x_3 - i\,(x_1 + 1))\,(x_0 + x_2 - x_3 + i\,(x_1 - 1)) \\ 2\,(x_1 - ix_2)\,(x_0 - x_3 - i) \end{bmatrix}, \tag{113}$$

where $\mathbf{F}$ is in a column vector notation. The general properties of the field (113) are discussed in [19, 20].

New Hopfions can be obtained from a given solution by performing infinitesimal conformal transformations, or a subgroup of them, on the functions

α and β. In order to see that, recall that the scalar functions change under the infinitesimal coordinate trasformations as follows [20]

$$x^{\mu} \rightarrow x'^{\mu} = x^{\mu} + \xi^{\mu} \tag{114}$$

$$\alpha(x) \rightarrow \alpha'(x') = \alpha(x) + \xi^{\nu}\partial_{\nu}\alpha(x) \,, \tag{115}$$

$$\beta(x) \rightarrow \beta'(x') = \beta(x) + \xi^{\nu}\partial_{\nu}\beta(x) \,, \tag{116}$$

where $\xi = \xi^{\mu}\partial_{\mu}$ is an arbitrary infinitesimal smooth vector field on $\mathbb{R}^{1,3}$. The equation (99) is invariant under the transformations (114) - (116) if the following condition is satisfied [20]

$$\epsilon_{rmn}\left[\delta_{jr}\left(\partial_0\xi_0 - \partial_s\xi_s\right) + i\epsilon_{jrs}\left(-\partial_0\xi_s + \partial_s\xi_0\right) + \partial_j\xi_r + \partial_r\xi_j\right]\partial_m\alpha\partial_n\beta = 0 \,. \tag{117}$$

The generators of the special conformal transformations are the vector fields

$$\xi_{\mu} = a_{\mu}\eta_{\mu\nu}x^{\mu\nu} - 2a_{\nu}x^{\nu}x^{\mu} \,. \tag{118}$$

It is easy to show that any vector field of the form (118) is a solution of the equation (117). By using this property of the special conformal transformations, the authors of [20] obtained new Bateman solutions characterized by integer powers p and q of the scalar functions α and β. The plane wave solution is one example of electromagnetic field from which new solutions can be generated by conformal deformations. The plane waves are characterized by the following scalar functions

$$\alpha = e^{\imath(x_3 - x_0)} \,, \qquad \beta = x_1 + ix_2. \tag{119}$$

The new functions that can be obtained from the equation (119) by conformal deformations have the following form

$$\alpha' = \exp\left[-1 + \frac{i\left(x_0 + x_3 - i\right)}{1 - x_0^2 + \delta_{mn}x^m x^n + 2ix_0}\right] , \tag{120}$$

$$\beta' = \frac{x_1 + ix_2}{1 - x_0^2 + \delta_{mn}x^m x^n + 2ix_0} \,. \tag{121}$$

The electric and magnetic fields built from α' and β' have *toric topology*. A generalization of this method to more complex knotted electromagnetic fields was given in [41]. The conformal deformations method was applied to the study of the physical properties of the optical vorticies in [42, 28].

5. KNOTS IN NONLINEAR ELECTRODYNAMICS

A very important application of Bateman method to non-linear electrodynamical models that generalize Maxwell's electrodynamics in the strong field regime, was given recently in [22, 23]. In these works, the authors showed that there are knot solutions of the equations of motion in any non-linear extension of Maxwell's electrodynamics that satisfies strong field requirement.

Let us see how the Bateman method can be applied to the Born-Infeld electrodynamics. The non-linear Born-Infeld action has the following form

$$S_{BI} = -\gamma^2 \int d^4x \left(\sqrt{1+F-P^2} - 1\right) , \tag{122}$$

where γ is a constant of dimension 2 and L and P are the usual Lorentz invariants defined as follows

$$L = \gamma^{-2}\delta_{mn}\left(B^m B^n - E^m E^n\right) , \tag{123}$$
$$P = \gamma^{-2}\delta_{mn}E^m B^n . \tag{124}$$

The scalars L and P are zero for all null-fields that satisfy the equations (50) and (51).

In order to prove that the theory described by the action S_{BI} has Hopfion solutions, the following argument has been developed in [22, 23]. Starting from the action (122), define new vector fields **H** and **D** whose components can be written in a notation analogous to Maxwell's electrodynamics

$$H_m = -\frac{\partial \mathcal{L}_{BI}}{\partial B^m} , \qquad D_m = \frac{\partial \mathcal{L}_{BI}}{\partial E^m} . \tag{125}$$

Here, $\mathcal{L}_{BI}$ is the Lagrangian density from the action (122). From it, one obtains the explicit form of H_m and D_m as follows

$$H_m = -\frac{1}{\sqrt{1+F-P^2}}\left(B_m - PE_m\right) , \tag{126}$$
$$D_m = -\frac{1}{\sqrt{1+F-P^2}}\left(E_m - PB_m\right) . \tag{127}$$

By applying the variational principle to the action (122), the following equations of motion are obtained

$$\partial_m D_m = 0\,, \quad (128)$$

$$\partial_m B_m = 0\,, \quad (129)$$

$$\varepsilon_{mnr}\partial_n E_r = -\partial_0 B_m\,, \quad (130)$$

$$\varepsilon_{mnr}\partial_n H_r = \partial_0 D_m\,. \quad (131)$$

The above equations show that the null fields from the relations (50) and (51) belong to a theory in which the equations of motion are of Maxwellian type. Therefore, by using the results presented in Section 2, one concludes that there are null field Hopfion solutions in the non-linear Born-Infeld electrodynamics.

The same argument can be generalized to all non-linear actions that in the weak field limit can be reduced to Maxwell's electrodynamics. Another interesting generalization is based on the observation that the equations of the classicall electrodynamics are similar to the equations of the fluid mechanics. Then by using the same reasoning as before, Hopfion solutions of the fluid flow lines were showed to exist in [22, 23].

Appendix - Hopf Map

In this Appendix, we briefly review the definition and some basic properties of the Hopf map that have been used in the text. A classical reference on this topic is the textbook [38] in which Hopf's theory is presented from a geometrical point of view.

Consider a n-dimensional unit sphere S^n defined as follows

$$S^n = \{\, \mathbf{x} \in R^{n+1} : \delta_{ij} x_i x_j = 1,\, i, j = 0, 1, \ldots, n \,\}\,. \quad (132)$$

The *Hopf fibration* is the mapping $h : S^3 \to S^2 \simeq \mathbb{CP}^1$ defined by the following *Hopf map*

$$h(x_0, x_1, x_2, x_3) = \left(x_0^2 + x_1^2 - x_2^2 - x_3^2\,, 2(x_0 x_3 + x_1 x_2)\,, 2(x_1 x_2 - x_0 x_3)\right)\,. \quad (133)$$

Alternatively, the Hopf map can be written in terms of the following complex coordinates on the spheres S^2 and S^3

$$S^2 = \{(x, z) \in \mathbb{R} \times \mathbb{C} : x^2 + |z|^2 = 1\}\,, \quad (134)$$

$$S^3 = \{(z_1, z_2) \in \mathbb{C}^2 : |z_1|^2 + |z_2|^2 = 1\}\,. \quad (135)$$

Then the Hopf map takes the following form

$$h(z_1, z_2) = \left(2z_1\bar{z}_2, |z_1|^2 - |z_2|^2\right) . \tag{136}$$

Since the projective space is a quotient set $\mathbb{CP}^1 \simeq S^3/U(1)$, the action of $U(1)$ on S^3 defines the fibres over S^3. These are mapped into the fibres over S^2 as follows. Consider the equivalence class of points of S^3 defined by the following relation

$$(z_1', z_2') \sim (z_1, z_2) \; : \; \text{if } \exists\, \lambda \in \mathbb{C} \text{ such that } (z_1', z_2') = (\lambda z_1, \lambda z_2) . \tag{137}$$

Then the Hopf map of (z_1', z_2') satisfies the following relation

$$h(z_1', z_2') = |\lambda|^2 h(z_1, z_2) . \tag{138}$$

The points (z_1', z_2') and (z_1, z_2) of S^3 belong to a fibre over S^2 if they are mapped onto the same point of S^2. This represents a constraint on the parameter λ which is satisfied by any λ of the form $\lambda = \exp(i\vartheta)$, where $\vartheta \in [0, 2\pi)$. Thus, λ necessarily belongs to the defining representation of $U(1)$.

From the geometrical point of view, the fibre of S^3 is the great circle that contains (z_1, z_2). An useful parametrization of fibres is given by the following relations

$$z_1 = \exp\left(i\xi + \frac{\varphi}{2}\right) \sin\left(\frac{\vartheta}{2}\right) , \qquad z_2 = \exp\left(i\xi - \frac{\varphi}{2}\right) \cos\left(\frac{\vartheta}{2}\right) . \tag{139}$$

It is easy to see that the Hopf map h takes the points from the fibres of S^3 to the following points on $S^2 \subset \mathbb{R}^3$

$$x_1 = 2|z_1||z_2| \cos(\varphi) , \tag{140}$$

$$x_2 = 2|z_1||z_2| \sin(\varphi) , \tag{141}$$

$$x_3 = |z_1|^2 - |z_2|^2 . \tag{142}$$

The stereographic projections can be easily determined from its definition and we find that there are several parametrizations of these projections. For example:

$$\pi_2(x_1, x_2, x_3) = \left(\frac{x_1}{1 - x_3}, \frac{x_2}{1 - x_3}\right) , \tag{143}$$

$$\pi_3(x_0, x_1, x_2, x_3) = \left(\frac{x_1}{1 - x_0}, \frac{x_2}{1 - x_0}, \frac{x_3}{1 - x_0}\right) . \tag{144}$$

The Hopf map inverse $\gamma = h^{-1}$ takes points (x, z) from S^2 into loops on S^3. In general, if the points are different $(x, z) \neq (x', z')$, so are the corresponding loops $\gamma \neq \gamma'$. However, there is an object associated to a pair of loops called the *Hopf invariant* of h that is the linking number of the pair of loops, also called the *Hopf index* of (γ, γ'), and which has the following general form

$$H(h) = l(\gamma, \gamma') \,. \tag{145}$$

$H(h)$ is an homotopy invariant that characterizes the Hopf map. It is useful to write the Hopf invariant in terms of the Chern-Simons integral [43] of some vector fields. This can be done as follows. Consider an unit vector field $\mathbf{U}(\mathbf{x})$ with the following properties

$$\delta_{ij} U_i(\mathbf{x}) U_j(\mathbf{x}) = 1 \,, \qquad |\mathbf{U}(\mathbf{x})| \to \mathbf{u} \text{ if } |\mathbf{x}| \to \infty \,, \tag{146}$$

where $\mathbf{u}$ is a constant unit vector. Then define the vector field $\mathbf{F}(\mathbf{x})$ with the following components

$$F_j(\mathbf{x}) = \varepsilon_{jmn} \varepsilon_{prs} U_p(\mathbf{x}) \partial_m U_r(\mathbf{x}) \partial_n U_s(\mathbf{x}) \,. \tag{147}$$

The field $\mathbf{F}(\mathbf{x})$ can be written in terms of a potential vector field $\mathbf{A}(\mathbf{x})$ as follows

$$F_j(\mathbf{x}) = \varepsilon_{jmn} \partial_m A_n(\mathbf{x}) \,. \tag{148}$$

Then the Hopf index of the field $\mathbf{U}(\mathbf{x})$ is given by the following equalities

$$H(\mathbf{U}) = \int d^3x \delta_{ij} F_i(\mathbf{x}) A_j(\mathbf{x}) = \int d^3x \varepsilon_{jmn} A_j(\mathbf{x}) \partial_m A_n(\mathbf{x}) \,. \tag{149}$$

We recognize in the last equality above the explicit form of the Chern-Simons integral for the potential $\mathbf{A}(\mathbf{x})$.

REFERENCES

[1] Trautman, A. (1977). "Solutions of the Maxwell and Yang-Mills Equations Associated with Hopf Fibrings," *Int. J. Theor. Phys.* 16, 561.

[2] Rañada, A. F. (1989). "A Topological Theory of the Electromagnetic Field," *Lett. Math. Phys.* 18, 97.

[3] Rañada, A. F. (1990). "Knotted solutions of the Maxwell equations in vacuum," *J. Phys. A* 23 L815.

[4] Rañada, A. F. (1992). "Topological electromagnetism," *J. Phys. A* 25, 1621.

[5] Rañada, A. F. and Trueba, J. L. (1995). "Electromagnetic Knots," *Phys. Lett. A* 202, 337-342.

[6] Rañada, A. F. and Trueba, J. L. (1997). "Two properties of electromagnetic knots," *Phys. Lett. A* 222, 25-33.

[7] Irvine, W. T. M. and Bouwmeester, D. (2008). "Linked and knotted beams of light," *Nature Physics* 4, 716 - 720.

[8] Irvine, W. T. M. (2010). "Linked and knotted beams of light, conservation of helicity and the flow of null electromagnetic fields," *J. Phys. A: Math. Theor.* 43 385203.

[9] Arrayás, M. and Trueba, J. L. (2010). "Motion of charged particles in a knotted electromagnetic field," *J. Phys. A* 43, 235401.

[10] Kleckner, D. and Irvine, W. T. M. (2013). "Creation and dynamics of knotted vortices," *Nature Physics* 9, 253-258.

[11] Arrayás, M. and Trueba, J. L. (2012). "Exchange of helicity in a knotted electromagnetic field," *Annalen Phys.* 524, 71.

[12] Arrayás, M. and Trueba, J. L. (2017). "Collision of two hopfions," *J. Phys. A* 50, 085203.

[13] Rañada, A. F., Tiemblo, A. and Trueba, J. L. (2017). "Time evolving potentials for electromagnetic knots," *Int. J. Geom. Meth. Mod. Phys.* 14, 1750073.

[14] Rañada, A. F. and Trueba, J. L. (1998). "A Topological mechanism of discretization for the electric charge," *Phys. Lett. B* 422, 196.

[15] Rañada, A. F. (2003). "Interplay of topology and quantization: topological energy quantization in a cavity," *Phys. Lett. A* 310, 434-444.

[16] Rañada, A. F. and Trueba, J. L. (2006). "Topological quantization of the magnetic flux," *Found. Phys.* 36, 427.

[17] Arrayás, M., Trueba, J. L. and Rañada, A. F. (2012). "Topological Electromagnetism: Knots and Quantization Rules," in *Trends in Electromagnetism - From Fundamentals to Applications*, Barsan, V. and Lungu, R. P. (Eds.), IntechOpen Ltd.

[18] Arrayás, M. and Trueba, J. L. (2011). "Electromagnetic Torus Knots," *J. Phys. A: Math. Theor.* 48, 025203.

[19] Kedia, H., Bialynicki-Birula, I., Peralta-Salas, D. and Irvine, W. T. M. (2013). "Tying knots in light fields," *Phys. Rev. Lett.* 111, 150404.

[20] Hoyos, C., Sircar, N. and Sonnenschein, J. (2015). "New knotted solutions of Maxwell's equations," *J. Phys. A* 48, no. 25, 255204.

[21] Goulart, É. (2016). "Nonlinear electrodynamics is skilled with knots," *Europhys. Lett.* 115, 10004.

[22] Alves, D. W.F., Hoyos, C., Nastase, H. and Sonnenschein, J. (2017). "Knotted solutions for linear and nonlinear theories: electromagnetism and fluid dynamics," *Phys. Lett. B* 773, 412.

[23] Alves, D. W. F., Hoyos, C., Nastase, H. and Sonnenschein, J. (2017). "Knotted solutions, from electromagnetism to fluid dynamics," *Int. J. Mod. Phys. A* 32, 1750200.

[24] Rañada, A. F. and Trueba, J. L. (1996). "Ball lightning an electromagnetic knot," *Nature* 383, 32.

[25] Irvine, W. T. M. and Kleckner, D. (2014). "Liquid crystals: Tangled loops and knots," *Nature Materials* 13, 229231.

[26] Smiet, C. B., Candelaresi, S., Thompson, A., Swearngin, J., Dalhuisen, J. W. and Bouwmeester, D. (2015). "Self-Organizing Knotted Magnetic Structures in Plasma," *Phys. Rev. Lett.* 115, 095001.

[27] Ji-Rong, R., Tao, Z. and Shu-Fan, M. (2008). J. R. Ren, T. Zhu and S. F. Mo, "Knotted topological phase singularities of electromagnetic field," *Commun. Theor. Phys.* 50, 1071.

[28] de Klerk, A. J. J. M., van der Veen, R. I., Dalhuisen, J. W. and Bouwmeester, D. (2017). "Knotted optical vortices in exact solutions to Maxwell's equations," *Phys. Rev. A* 95, 053820.

[29] Trueba, J. L. (2008). "Electromagnetic knots and the magnetic flux in superconductors," *Ann. Fond. Louis de Broglie* 33, 183-192.

[30] Arrayás, M., Bouwmeester, D. and Trueba, J. L. (2017). "Knots in electromagnetism," *Phys. Rept.* 667.

[31] Kopiński, J. and Natário, J. (2017). "On a remarkable electromagnetic field in the Einstein Universe," *Gen. Rel. Grav.* 49, 81.

[32] Vancea, I. V. (2017). "On the existence of the field line solutions of the Einstein - Maxwell equations," *Int. J. Geom. Meth. Mod. Phys.* 15, 1850054.

[33] Costa e Silva, W., Goulart, E. and Ottoni, J. E. (2018). "On spacetime foliations and electromagnetic knots," arXiv:1809.09259 [math-ph].

[34] Alves, D. W. F. and Nastase, H. (2018). "Hopfion solutions in gravity and a null fluid/gravity conjecture," arXiv:1812.08630 [hep-th].

[35] Vancea, I. V. (2019). "Field Line Solutions of the Einstein-Maxwell Equations," chapter in this volume.

[36] Jackson, J. D. (1962). "Classical Electrodynamics," John Wiley and Sons.

[37] Frankel, T. (1997). "The geometry of physics: An introduction," Cambridge University Press.

[38] Milnor, J. (1997). "Topology from the Differentiable Viewpoint," Princeton University Press.

[39] Bateman, H. (2016). "The Mathematical Analysis of Electrical and Optical Wave-Motion," Cambridge University Press.

[40] Thompson, A., Wickes, A., Swearngin, J. and Bouwmeester, D. (2015). "Classification of Electromagnetic and Gravitational Hopfions by Algebraic Type," *J. Phys. A* 48, 205202.

[41] Kedia, H., Foster, D., Dennis, M. R. and Irvine, W. T. M. (2016). "Weaving knotted vector fields with tunable helicity," *Phys. Rev. Lett.* 117, 274501.

[42] Dennis, M. R., King, R. P., Jack, B., O'Holleran, K. and Padgett , M. J. (2010). "Isolated optical vortex knots," *Nature Physics* 6, 118-121.

[43] Whitehead, J. H. C. 1947. "An expression of Hop's invariant as an integral," *Proc. Nat. Acad. Sci. U.S.A.* 33, 117123.

In: An Essential Guide to Maxwells Equations ISBN: 978-1-53616-680-4
Editor: Casey Erickson

Chapter 2

FIELD LINE SOLUTIONS OF THE EINSTEIN-MAXWELL EQUATIONS

Ion V. Vancea*
Grupo de Física Teórica e Matemática Física
Departamento de Física
Universidade Federal Rural do Rio de Janeiro
Seropédica - Rio de Janeiro, Brazil

Abstract

In this paper, we are going to review the gravitating electromagnetic field in the 1+3 formalism on a general hyperbolic space-time manifold. We also discuss the recent results on the existence of the local field line solutions of the Einstein-Maxwell equations that generalize the Rañada solutions from the flat space-time. The global field line solutions do not always exist since the space-time manifold could impose obstructions to the global extension of various geometric objects necessary to build the fields. One example of a gravitating field line solution is the Kopiński-Natário field which is discussed in some detail.

Keywords: Einstein-Maxwell equations, local field line solutions, generalization of Rañada solutions

---

*Corresponding Author's E-mail: ionvancea@ufrrj.br.

1. INTRODUCTION

Recently, there has been an increasing interest in the topological aspects of the electromagnetism. This research line was pioneered by Trautman and Rañada in the seminal papers [1, 2, 3]. Since then, many developments and applications of the topological electromagnetic fields have appeared in the literature, ranging from the atomic particles physics to the colloidal matter physics. The convergence of classical electrodynamics and topology is mutually beneficial for both fields. For a recent list of references and an updated review of the topological solutions in the classical electrodynamics, we refer to [4, 5].

Most of the research on the topological electromagnetic fields has been done in the context of the classical electrodynamics applied to low energy phenomena or in the vacuum. However, the electromagnetic fields have a larger range of applications which include cosmological and astrophysical processes that usually take place at high energy. Generalizing the topological electromagnetism to these systems poses new challenges from the physical and mathematical point of view since in these cases one has to deal with the gravitating electromagnetic fields. Very recently, important steps in analyzing the topological properties of gravitons as well as of higher spin particles have been taken in [6, 7, 8, 9, 10]. These works are important for the topological electromagnetism since not only they generalize the topological properties of the classical photons but also contain new information on spin-2 field viewed as a particular case of the higher spin fields. However, the analyses from [6, 7, 8, 9, 10] focus on spin-2 fields in flat space-time. Therefore, they do not address the problem of gravitating topological electromagnetism. That problem is studied in [11], where it was proved that there are local field line solutions of Einstein-Maxwell equations on hyperbolic space-time manifolds that are a direct generalization of Rañada's solutions. Another contribution to the gravitating topological electromagnetism can be found in [12], where a particular topological solution of the Einstein-Maxwell equations in the Einstein universe was given. Also, it is worthwhile mentioning the work [13], in which the electromagnetic knots on hyperbolic manifolds are treated formally from the point of view of the foliation theory.

In the present chapter, we are going to give a pedagogical introduction to the problem of the topological gravitating electromagnetic fields. Since the space-time assumes the properties of a curved manifold in the presence of gravity, there are difficulties in analysing the gravitating fields inherited from the general formulation of the field theories on curved manifolds. While in the flat space-time, the topological electromagnetic solutions of the linear Maxwell equations can be calculated in the vacuum, in the presence of the gravitational field, the vacuum contains components of the metric that do not decouple from the electromagnetic field. In this case, the topological fields like knots and tori are better described in terms of electric and magnetic field lines which make more transparent their geometrical and topological properties. However, the decomposition of the electromagnetic field in electric and magnetic components is not possible on general curved manifolds. One remarkable exception are the hyperbolic manifolds for which there is a canonical formulation of the General Relativity, called *1+3 - formalism*, that allows one to put the Einstein-Maxwell equation in a form similar to the Maxwell equations in the Minkowski space-time [14]. By using the $1 + 3$ - formalism, it has been shown in [11] that Rañada's solutions have a *local generalization* to hyperbolic manifolds. However, one cannot expect *global solutions* to exist on general hyperbolic manifolds. Indeed, global field line solutions depend on the global properties of the space-time manifold that could contain obstructions to the global extension of certain mathematical objects such as the differential forms. Therefore, the global solutions are remarkable and they are associated to particular manifolds. This is the case of the static Einstein universe for which a topological radiation field that depends only on time was found in [12].

This work is organized as follows. In Section 2, we are going to review the basic features of the Maxwell equations in flat space-time in terms of differential forms. Since our main focus is the generalization of Rañada's solutions to the gravitating electromagnetic field, we will introduce in Section 3 the $1 + 3$ - formalism which makes the correspondence between the equations of motion of the electromagnetic field in flat and in curved space-time, respectively, more transparent. This formalism will be used to write the Einstein-Maxwell equations on hyperbolic manifolds. In Section 4, we will revisit the proof of existence of local field line solutions of Einstein-Maxwell equations on hyperbolic manifolds from [11]. In Section 5, the Kopiński-Natário solution in the Einstein space is presented. We collect some useful mathematical definitions in the Appendix. The units used throughout this paper are natural with $c = G = 1$.

2. Field Line Solutions in Flat Space-Time

In this section, we are going to survey the Maxwell equations in the Minkowski space-time and establish our notations. The material presented here is well-known and can be found in standard texts on classical electrodynamics, see e. g. [16]. We also review the Rañada field line solutions from [2, 3]. A more detailed review can be found in a different chapter from this volume [5].

2.1. The Maxwell's Equations

Consider the Minkowski space-time $\mathbb{R}^{1,3} = \{\mathbb{R}^4, \eta\}$, where η is the Minkowski pseudo-metric tensor of signature $(-,+,+,+)$. The events from $\mathbb{R}^{1,3}$ are identified by the coordinate four-vectors $x^\mu = (t, \mathbf{x}) = (t, x^i)$, where the indices $i, j, k = 1, 2, 3$ denote the space-like components. The electromagnetic field on $\mathbb{R}^{1,3}$ can be described in terms of the four-vector potential $A^\mu = (\phi, \mathbf{A})$, where ϕ is the scalar potential and $\mathbf{A} = (A^1, A^2, A^3)$ is the three-dimensional vector potential. Then the electromagnetic field is given by the rank-2 antisymmetric tensor $F_{\mu\nu}$ defined by the following relation

$$F_{\mu\nu} = \partial_\mu A_\nu - \partial_\nu A_\mu \,. \tag{1}$$

The tensor $F_{\mu\nu}$ has six independent degrees of freedom which are identified with the components of the electric and magnetic vector fields as follows

$$E_i = \partial_i A_0 - \partial_0 A_i \,, \qquad B_i = \varepsilon_{ijk}\partial_j A_k \,. \tag{2}$$

The Hodge dual (pseudo)-tensor associated to $F_{\mu\nu}$ is defined by the following relation

$$\star F_{\mu\nu} = \frac{1}{2}\varepsilon_{\mu\nu\rho\sigma}F^{\rho\sigma} \,, \qquad F^{\mu\nu} = \eta^{\mu\rho}\eta^{\nu\sigma}F_{\rho\sigma} \,. \tag{3}$$

One can write the electromagnetic field in terms of differential forms on $\mathbb{R}^{1,3}$. This formulation will be useful later when we will discuss the Einstein-Maxwell equations. Let us introduce the 2-form electromagnetic field F in the three-dimensional and four-dimensional representations, respectively, given by the following relations

$$F = B + E \wedge dx^0 \,, \tag{4}$$

$$F = F_{\mu\nu}\, dx^\mu \wedge dx^\nu \,, \tag{5}$$

where B is the magnetic 2-form, E is the electric 1-form and $\wedge$ is the wedge product of differential forms. Then the Hodge star operator is defined by the following relations

$$\star : \Omega^k(\mathbb{R}^{1,3}) \to \Omega^{4-k}(\mathbb{R}^{1,3}) \,, \tag{6}$$

$$\omega \wedge (\star\sigma) = \langle \omega, \sigma \rangle \mathrm{n} \,, \tag{7}$$

where n is an unitary vector and $\langle \cdot, \cdot \rangle$ is the scalar product. Here, we denote by $\Omega^k(M)$ the linear space of k-forms on the manifold M. The Maxwell equations in terms of differential forms are given by just two differential equations

$$dF = 0\,, \tag{8}$$

$$\star\, d \star F = J\,, \tag{9}$$

where $J = J_\mu dx^\mu$ is the 1-form current density [17]. The equations (8) and (9) are covariant which means that their symmetry group is the Poincaré group.

In what follows, we are going to discuss a particular type of solutions of Maxwell's equations in the vacuum which is defined by the absence of currents, namely,

$$dF = 0\,, \tag{10}$$

$$\star\, d \star F = 0\,, \tag{11}$$

In this case, the Maxwell equations have an additional symmetry which is the electromagnetic duality given by the interchange of the forms F and $\star F$ with each other

$$F \leftrightarrow \star F\,. \tag{12}$$

It is important to recall a property of the Hodge star operator. If ω is an arbitrary k-dimensional form on a manifold M of dimension n, then

$$\star\star\, \omega = (-1)^{k(n-k)} \text{sign}\,(\det \eta)\, \omega\,. \tag{13}$$

It follows from the equation (13) that if $M = \mathbb{R}^{1,3}$ and $k = 2$, the Hodge star operator satisfies the equation

$$\star^2 = -1\,. \tag{14}$$

From the equation (14), one can deduce that the operator $\star^2$ has the eigenvalues $\pm i$ when acting on the 2-forms. Therefore, the fields F belong to one of the two classes of self-dual or anti-self-dual 2-forms. Due to the linearity of Maxwell's equations (10) and (11), a general solution in vacuum is a superposition of solutions from each class

$$F = F_+ + F_-\,, \qquad \star F_\pm = \pm i F\,. \tag{15}$$

The above equations shows that F is a complex 2-form. It is a simple exercise in electrodynamics to prove that the equations (10) and (11) reproduce the three-dimensional Maxwell equations. This can be seen by decomposing the differential forms defined on the Minkowski space-time with respect to its global spatial foliation whose leaves are isomorphic to $\mathbb{R}^3$. The derived electromagnetic forms dF and $\star d \star F$ have the following decomposition

$$dF = dB + dE \wedge dx^0 \,, \tag{16}$$

$$\star d \star F = -\partial_0 E - \star \mathbf{d} \star E \wedge dx^0 + \star \mathbf{d} \star B \,, \tag{17}$$

where $\mathbf{d}$ is the three-dimensional differential derivative and $\star$ is the three-dimensional Hodge star operator. Then the Maxwell equations take the following form

$$\mathbf{d}E + \partial_0 B = 0 \,, \tag{18}$$

$$\mathbf{d}B = 0 \,, \tag{19}$$

$$\star \mathbf{d} \star E = \rho \,, \tag{20}$$

$$\star \mathbf{d} \star B - \partial_0 E = \mathbf{J} \,, \tag{21}$$

where $\mathbf{J} = J_i dx^i$. In the vacuum, the equations (21) have a simpler form

$$\mathbf{d}E + \partial_0 B = 0 \,, \tag{22}$$

$$\mathbf{d}B = 0 \,, \tag{23}$$

$$\mathbf{d} \star E = 0 \,, \tag{24}$$

$$\star \mathbf{d} \star B - \partial_0 E = 0 \,. \tag{25}$$

The formalism of differential forms allows one to calculate the properties of the electromagnetic field in a coordinate independent manner. Also, the equations obtained in this formalism are more compact. For more details we refer the reader to the chapter [5] from this volume or to the excellent book [17].

2.2. Field Line Solutions

In order to describe the dynamics of the electromagnetic field in the vacuum, the equations (22) - (25) must be solved with proper boundary conditions. As it is well known, the most general solution to the above set of equations is given by a superposition of monocromatic waves that satisfy the dispersion relation

$\mathbf{k}^2 = \omega^2(\mathbf{k})$ where $\mathbf{k}$ is the wave vector and $\omega(\mathbf{k})$ is the correspondig wave frequency in the infinite empty space $\mathbb{R}^3$ [16]. Since these solutions are known for a long time, it came as a surprise when Trautman and Rañada published three articles in which they presented independetly of each other, a new type of solution of Maxwell's equations in vacuum with a non-trivial topological structure [1, 2, 3]. All solutions of the topological class are characterized by new non-zero *topological charges* which are the *link numbers* between the electric and magnetic field lines. Also, the topological solutions can be expressed in terms of field lines which justifies their name of *field line solutions*. A more extended review of the field line solutions is presented in the chapter [5] from this volume. Here, we are only going to recall the general form of Rañada's solutions which is relevant for the rest of our discussion.

The Rañada solutions are described in terms of two smooth complex scalar fields ϕ and θ in the Minkwoski space-time [4]

$$\phi : \mathbb{R}^{1,3} \to \mathbb{C}\,, \qquad \theta : \mathbb{R}^{1,3} \to \mathbb{C}\,. \tag{26}$$

The main role of these fields is to serve as a backbone for the topology of the field lines in $\mathbb{R}^3$ in the following sense: the electric and magnetic field lines are the level curves of ϕ and θ, respectively. Then one can show that all solutions of Maxwell's equations that have this property are of the following form

$$F_{\mu\nu} = g(\bar{\phi}, \phi)\left(\partial_\mu \bar{\phi}\, \partial_\nu \phi - \partial_\nu \bar{\phi}\, \partial_\mu \phi\right)\,, \tag{27}$$

$$\star F_{\mu\nu} = f(\bar{\theta}, \theta)\left(\partial_\mu \bar{\theta}\, \partial_\nu \theta - \partial_\nu \bar{\theta}\, \partial_\mu \theta\right)\,. \tag{28}$$

The field line solutions given by the equations (27) and (28) are parametrized by two smooth function g and f that depend on θ and ϕ. The electromagnetic differential forms are related to the tensors $F_{\mu\nu}$ and $\star F_{\mu\nu}$ by the following relations

$$F = -\varepsilon_{jkl} B_j dx^k \wedge dx^l + E_j dx^j \wedge dx^0\,, \tag{29}$$

$$\star F = \varepsilon_{jkl} E_j dx^k \wedge dx^l + B_j dx^j \wedge dx^0\,. \tag{30}$$

A Rañada solution of Maxwell's equations is an electromagnetic field of the form (27) and (28) that has the following components

$$E_j = \frac{\sqrt{a}}{2\pi i}\left(1 + |\theta|^2\right)^{-2} \varepsilon_{jkl}\, \partial_k \bar{\theta}\, \partial_l \theta\,, \tag{31}$$

$$B_j = \frac{\sqrt{a}}{2\pi i}\left(1 + |\phi|^2\right)^{-2} \varepsilon_{jkl}\, \partial_k \bar{\phi}\, \partial_l \phi\,. \tag{32}$$

Here, a is a parameter that fixes the physical dimension of the solution. The fields $\mathbf{E}$ and $\mathbf{B}$ defined by the equations (31) and (32) correspond to particular choices of f and g from the equations (27) and (28). Also, we obtain the following constraints on the paramenters from the self-duality condition in three-dimensions [4]

$$\left(1+|\phi|^2\right)^{-2}\varepsilon_{jmn}\partial_m\phi\partial_n\bar{\phi} = \left(1+|\theta|^2\right)^{-2}\left(\partial_0\bar{\theta}\partial_j\theta - \partial_0\theta\partial_j\bar{\theta}\right), \tag{33}$$

$$\left(1+|\theta|^2\right)^{-2}\varepsilon_{jmn}\partial_m\bar{\theta}\partial_n\theta = \left(1+|\phi|^2\right)^{-2}\left(\partial_0\bar{\phi}\partial_j\phi - \partial_0\phi\partial_j\bar{\phi}\right). \tag{34}$$

Some comments are in order here. Firstly, note that the Rañada solutions satisfy the orthogonality property

$$E_j B_k \delta_{jk} = 0\,. \tag{35}$$

Secondly, since the fields given by the relations (27) and (28) satisfy the Maxwell equations, the charges associated to the Poincaré group and the $U(1)$ group are conserved. However, there are new charges associated to the topology of the field lines. The topological observables related to the topological charges are the pure electric and magnetic helicities of the electromagnetic field that are defined as follows

$$H_{ee} = \int d^3x\, \delta_{ij} E_i C_j = \int d^3x\, \varepsilon_{jkl} C_j \partial_k C_l\,, \tag{36}$$

$$H_{mm} = \int d^3x\, \delta_{ij} B_i A_j = \int d^3x\, \varepsilon_{jkl} A_j \partial_k A_l\,, \tag{37}$$

where $\mathbf{A}$ and $\mathbf{C}$ are the corresponding magnetic and electric potential vectors

$$E_j = \varepsilon_{jkl}\partial_k C_l\,, \qquad B_j = \varepsilon_{jkl}\partial_k A_l\,. \tag{38}$$

Observed that the correspondence between the helicities and the topological objects is a consequence of the definition of the field line solution in terms of the complex scalar fields ϕ and θ.

If the electromagnetic solutions carry a finite energy, then they should take zero value in the limit $|\mathbf{x}| \rightarrow \infty$. In order for that to happen, the fields $\phi(t,\mathbf{x})$ and $\theta(t,\mathbf{x})$ must have the same asymptotic behaviour as the energy. That implies that ϕ and θ are complex functions on $\mathbb{R}\times S^3$ or, equivalently, they correspond to one-parameter families of maps $S^3 \rightarrow \mathbb{C}$. On the other hand, there is a natural identification $\mathbb{C} \simeq \mathbb{R}^2$. This identification can be further refined if the inverse maps ϕ^{-1} and θ^{-1} do not depend on the complex phases of their arguments. The functions that satisfy this property can be interpreted as maps $S^3 \to S^2$. However, these are the Hopf maps characterized by the topological number called the *Hopf index*. It turns out that when this index is expressed as a Chern-Simons integral, it takes the same form as the heliticies defined above. For a detailed discussion of these aspects, see [4].

3. THE $1+3$ - FORMALISM OF THE GENERAL RELATIVITY

In this section, we review those basic concepts of the $1 + 3$ - formalism of the General Relativity necessary to discuss the Einstein-Maxwell equations on hyperbolic manifolds. In our presentation, we will follow mainly the reference [14] with minor modifications of the notations.

3.1. Space-Time Foliation

According to the General Relativity, the gravity is a consequence of the non-trivial geometry of the space-time viewed as a four-dimensional differential manifold M endowed with a metric tensor field $\mathbf{g} = g_{\mu\nu}$. The manifold M can be approximated locally by the Minkowski space-time [15]. It follows from the first principles that the metric $g_{\mu\nu}$ must be a solution of the Einstein equations either in the presence of the matter or with no matter at all. In this case, the dynamics of the gravitating electromagnetic field is given by a set of covariant equations on M. Our main goal is to discuss the existence of field line solutions of these equations. More concretely, we would like to see whether there are any local solutions of the equations of motion of the gravitating electromagnetic field that generalize the Rañada solutions from the flat space-time. The existence problem is not well posed globally unless further assumptions about the structure of the hyperbolic manifold are made. In the general case, there could be obstructions to the existence of certain mathematical objects, such as the 2-forms, that could prevent the existence of the global fields on M.

We recall here two important properties of the mathematical structure of Maxwell's equations that play a silent role in the derivation of the results from the previous section: *i*) the splitting between the electric and the magnetic fields, necessary to define the electric and magnetic field lines; and *ii*) the explicit time-evolution of the system during which the helicities are conserved. These elements can be reproduced in the presence of gravity if the manifold M is globally hyperbolic, that is if it admits a foliation in terms of an one-parameter family of space-like manifolds $\Sigma \simeq \mathbb{R}^3$ parametrized by a global time $t \in \mathbb{R}$. In that case, the manifold M has the topology $M = \mathbb{R} \times \Sigma$. The method of splitting the covariant equations on M with respect to the foliation is called the *1+3 - formalism.* In order to make the discussion more concrete, we need to introduce these concepts in a more formal way. For more details on the 1+3 - formalism see [14].

Let $(M, \mathbf{g})$ denote a four-dimensional space-time manifold M endowed with a smooth metric tensor field $\mathbf{g}$ of signature $(-,+,+,+)$ that has the following properties:

- M is time-orientable;
- M is hyperbolic.

Then there is a globally defined scalar field $\mathbf{t}$ such that M is a foliation generated by $\mathbf{t}$ with the leaves defined by the following relation

$$\Sigma_t = \{p \in \mathcal{M} : \mathbf{t}(p) = t = \text{constant}\} . \tag{39}$$

Equivalently, one can write the foliation as $\mathcal{M} \simeq \mathbb{R} \times \Sigma$.

Among the mathematical structures that can be defined on M, there are two remarkable vector fields and one scalar function that play a crucial role in the $1+3$-formalism, namely: the normal vector field $\mathbf{n}$, the normal evolution vector field $\mathbf{m}$ and the lapse function N. These are defined by the following relations

$$\mathbf{n} := -N\nabla\mathbf{t} , \qquad \mathbf{m} := N\mathbf{n} , \qquad N := \left[-g_{\mu\nu}\nabla^{\mu}\mathbf{t}\nabla^{\nu}\mathbf{t}\right]^{-\frac{1}{2}} . \tag{40}$$

Since we want to identify the leaves Σ_t at every value of the parameter t and at every point p with the space-like submanifold of M, and we want the vector field $\mathbf{t}$ to be the time vector field throughout the entire M, the gradient $\nabla\mathbf{t}$ must be time-like.

In order to be able to make physical measurements at a point $p \in \Sigma_t$ one needs local coordinates. One possibility is to choose $x^{\mu} = (t, x^i)$ adapted to the foliation. Here, the Latin indices $i, j = 1, 2, 3$ are for the components on the leaf. The time evolution of a physical system that contains the event p at t takes place along the field line of the vector field $\boldsymbol{\partial}_t$ defined by the following relations

$$\boldsymbol{\partial}_t := \mathbf{m} + \boldsymbol{\beta}, \quad g_{\mu\nu}\beta^{\mu}n^{\nu} = 0, \tag{41}$$

where $\boldsymbol{\partial}_t$ is the derivative along the adapted time and $\boldsymbol{\beta} \in \mathcal{T}_p(\mathcal{M})$ is the shift vector corresponding to the displacement of the origin of the space-like coordinates between two infinitesimally closed leaves.

The above decomposition of the manifold M is in agreement with its topological structure as a foliation. All geometrical objects defined on M can be decomposed in the same way. For example, the four-dimensional metric tensor g on M induces a three-dimensional metric $\boldsymbol{\gamma}$ on each leaf Σ_t by the following canonical reduction operation

$$\boldsymbol{\gamma} = \mathbf{g}|_{\Sigma_t} \Longleftrightarrow \gamma_{ij} = g_{ij} \, . \tag{42}$$

The equation (42) can be interpreted as the action of a projector $P^{\mu}{}_{\nu}$ on the tensor g with

$$P^{\mu}{}_{\nu} := g^{\mu}{}_{\nu} + n^{\mu} n_{\nu} \, , \qquad P_{\mu i} \, n^{\mu} = 0 \, . \tag{43}$$

This second interpretation is more operational as it allows one to project other mathematical objects onto Σ_t.

If we denote the components of the covariant derivative on the leave by D_i, the three-dimensional connection associated to it is torsionless providing that the metric is compatible with the covariant derivative $D^i \gamma_{ij} = 0$. Since the leaf Σ_t is embedded into the four-dimensional manifold M, one can define its exterior curvature as follows

$$K_{ij} := P^{\mu}{}_{i} P^{\mu}{}_{j} \nabla_{\mu} n_{\nu} \, , \qquad K := \gamma^{ij} K_{ij} \, , \tag{44}$$

where ∇_{μ} are the components of the four-dimensional covariant derivative compatible with the metric $g_{\mu\nu}$. It is easy to see that the line element takes the following form under the metric decomposition from the equation (42)

$$ds^2 = -(N dt)^2 + \gamma_{ij} \left(dx^i + \beta^i dt \right) \left(dx^j + \beta^j dt \right) \, . \tag{45}$$

By acting with the projector $P^{\mu}{}_{\nu}$, one can decompose the covariant gravitational field and the electromagnetic field with respect to the foliation of the underlying manifold.

3.2. Einstein-Maxwell Equations

The equations that govern the dynamics of the gravitating electromagnetic field can be obtained from the following action

$$S[g, A] = - \int d^4 x \sqrt{-g} \left(F_{\mu\nu} \, F^{\mu\nu} + A_{\mu} \, j^{\mu} \right) \, , \tag{46}$$

where

$$F_{\mu\nu} = \partial_\mu A_\nu - \partial_\nu A_\mu \,, \tag{47}$$

$$j^\mu = \frac{1}{\sqrt{-g}} \partial_\nu \mathcal{D}^{\mu\nu} \,, \tag{48}$$

$$\mathcal{D}^{\mu\nu} = \sqrt{-g} F^{\mu\nu} \,. \tag{49}$$

By applying the variational principle to the action (46), one obtains the equations of motion of the electromagnetic field as well as of the gravitational field. These equations are the Einstein-Maxwell equations. In the $1 + 3$ - formalism presented above, the field strength tensor $F_{\mu\nu}(t, \mathbf{x})$ can be decomposed locally in to the electric field $E^\mu(t, \mathbf{x})$ and the magnetic field $B^\mu(t, \mathbf{x})$, respectively. The components of the electromagnetic tensor are given by the following relations

$$E_\mu(t, \mathbf{x}) = F_{\mu\nu}(t, \mathbf{x}) n^\nu(t, \mathbf{x}) \,, \quad B_\mu(t, \mathbf{x}) = \frac{1}{2} \varepsilon_{\mu\nu\sigma}(t, \mathbf{x}) F^{\nu\sigma}(t, \mathbf{x}) \,, \tag{50}$$

where $\varepsilon_{\mu\nu\sigma}(t, \mathbf{x})$ is the contracted four-dimensional Levi-Civita tensor. One can easily show that the electric and magnetic fields satisfy the following equations

$$E_\mu(t, \mathbf{x}) n^\mu(t, \mathbf{x}) = 0 \,, \quad B_\mu(t, \mathbf{x}) n^\mu(t, \mathbf{x}) = 0 \,. \tag{51}$$

The equations (51) show that $E^\mu(t, \mathbf{x}) \,, B^\mu(t, \mathbf{x}) \in \mathcal{T}_p(\Sigma_t)$, that is the electric and magnetic vectors are tangent to the leaf at p. It follows from (51) that the field strength has the following local form

$$F_{\mu\nu} = n_\mu E_\nu - n_\nu E_\mu + \varepsilon_{\mu\nu\rho\sigma} n^\rho B^\sigma \,. \tag{52}$$

In what follows, we are going to discuss some properties of the Einstein-Maxwell equations. In order to simplify the notation, we will drop off the local space-time coordinates unless their explicit presence is strictly necessary. In this notation, the Einstein-Maxwell equations without sources are written as

$$\mathcal{L}_{\mathbf{m}} E^i - NKE^i - \varepsilon^{ijk} D_j \left(N B_k \right) = 0, \tag{53}$$

$$\mathcal{L}_{\mathbf{m}} B^i - NKB^i + \varepsilon^{ijk} D_j \left(N E_k \right) = 0, \tag{54}$$

$$D_i E^i = 0, \tag{55}$$

$$D_i B^i = 0. \tag{56}$$

Note that the equations (53) and (54) describe the dynamics of the electromagnetic field and they correspond to the Faraday and Ampère laws, respectively. The equations (55) and (56) are constraints on the components E^i and B^i and generalize the Gauss law to the gravitating electromagnetic field.

4. Existence of Local Field Line Solutions

The 1+3 - formalism introduced in the previous section allows us to separate the covariant electromagnetic field in to electric and the magnetic components. This decomposition is useful if one wants to generalize the field line solutions from the Minkowski space-time to hyperbolic manifolds. In the most general case of an arbitrary hyperbolic space-time manifold (M, g), the field line solutions are local. Nevertheless, global field line solutions could exist in particular case. In the rest of this section, we will review the arguments from [11] where it was given the proof of existence of local field line solutions of Einstein-Maxwell equations on general hyperbolic manifolds.

4.1. Field Line Solutions

In order to find magnetic field line solutions of the Einstein-Maxwell equations, we need to analyse only the equations (54) and (56). Since the problem is local, one must work within an (arbitrary) neighbourhood $U_p \in \mathcal{M}$, where $p \in \Sigma_t$. Also, it is necessary to use the adapted coordinates $(t, \mathbf{x})$ in U_p as discussed before.

As we have seen in the previous section, the magnetic field lines are defined in terms of scalar fields in the Minkowski space-time. That suggests that a scalar field $\phi : U_p \to \mathbb{C}$ be introduced on $U_p \in M$. The equation (54) is the equation of motion of $B^i(t, \mathbf{x})$ which should also obey the constraint (56) at all times. The form of this constraint suggests the following ansatz for the magnetic field

$$B^i(t, \mathbf{x}) = f(t, \mathbf{x})\varepsilon^{ijk}(t, \mathbf{x})D_j\phi(t, \mathbf{x})D_k\bar{\phi}(t, \mathbf{x}). \tag{57}$$

Here, $f(t, \mathbf{x})$ is a smooth arbitrary field on U_p. By plugging $B^i(t, \mathbf{x})$ into the equation (56), we can verify that any function $f(t, \mathbf{x})$ that depends on the space-time coordinates only implicitly via the scalar field, i. e. $f(\phi(t, \mathbf{x}), \bar{\phi}(t, \mathbf{x}))$, satisfies the ansatz (57).

Let us look at the electric component corresponding to the magnetic field (57). The Ampère law (54) determines the evolution of $B^i(t, \mathbf{x})$ in terms of metric and of components $E^i(t,\mathbf{x})$ of the electric field. By contemplating the equation (54) and the ansatz (57), we conclude that the electric field should have the following form

$$E^i(t,\mathbf{x}) = \frac{f(t,\mathbf{x})}{N(t,\mathbf{x})}\left[\left(\mathcal{L}_{\mathbf{m}}\bar{\phi}(t,\mathbf{x})\right)D^i\phi(t,\mathbf{x}) - \left(\mathcal{L}_{\mathbf{m}}\phi(t,\mathbf{x})\right)D^i\bar{\phi}(t,\mathbf{x})\right]. \tag{58}$$

The fields proposed in the equations (57) and (58) must verify the equation of motion (54). The verification can be done by plugging the fields into the Ampère law. Then the most rapid way to prove that the equation (54) is satisfied is to show that both left- and right-hand sides of it take the same form, namely,

$$\begin{aligned}&\varepsilon^{ijk} f\left(\partial_t\partial_j\phi - \beta^r\partial_r\partial_j\phi - \partial_r\phi\partial_j\beta^r\right)\partial_k\bar{\phi}\\ &+\varepsilon^{ijk} f\left(\partial_t\partial_k\bar{\phi} - \beta^r\partial_r\partial_k\bar{\phi} - \partial_r\bar{\phi}\partial_k\beta^r\right)\partial_j\phi.\end{aligned} \tag{59}$$

This result can be proved by direct calculations on each side of the equation. For more details we refer to [11].

The above analysis lead us to the conclusion that the fields $B^i(t,\mathbf{x})$ and $E^i(t,\mathbf{x})$ given by the relations (57) and (58) are field line solutions of two of the Einstein-Maxwell equations. These solutions satisfy the orthogonality property

$$\gamma^{ij}(t,\mathbf{x})E_i(t,\mathbf{x})B_j(t,\mathbf{x}) = 0\,, \qquad \forall p \in U_p\,. \tag{60}$$

As in the flat space-time, the magnetic field lines are the level lines of the complex function ϕ. However, the interpretation of the electric field in terms of field lines is not transparent in the relation (58). That can be remedied by constructing an electric field line solution as was done in the flat space-time. An important tool that was used there was the invariance of Maxwell's equations under the duality between the electric and magnetic fields in the vacuum. In the absence of sources, the Einstein-Maxwell equations without sources are invariant under the electric-magnetic duality, too, as it was shown in [18]. This symmetry allows one to find the line solutions of the Faraday law and the electric Gauss law which are given by the equations (53) and (55), respectively.

Let us introduce a second complex field $\theta : U_p \to \mathbb{C}$ and note that the constraint on the electric field is the same as the one on the magnetic field. It follows that one can take the electric field of the same form as the magnetic field given by the equation (57). Then we can write

$$E^i(t,\mathbf{x}) = g(t,\mathbf{x})\varepsilon^{ijk}(t,\mathbf{x})D_j\bar{\theta}(t,\mathbf{x})D_k\theta(t,\mathbf{x}), \tag{61}$$

where $g(t, \mathbf{x})$ is an arbitrary real smooth field on U_p that depends on the adapted coordinates only implicitly as $g(\theta(t, \mathbf{x}), \bar{\theta}(t, \mathbf{x}))$. By repeating the arguments given above where we have discussed the electric field line solution, we conclude that the magnetic field line should have the following form

$$B^i(t, \mathbf{x}) = \frac{g(t, \mathbf{x})}{N(t, \mathbf{x})} \left[\left(\mathcal{L}_{\mathbf{m}} \bar{\theta}(t, \mathbf{x}) \right) D^i \theta(t, \mathbf{x}) - \left(\mathcal{L}_{\mathbf{m}} \theta(t, \mathbf{x}) \right) D^i \bar{\theta}(t, \mathbf{x}) \right]. \quad (62)$$

Since the equation of motion and the constraints for the electric and magnetic fields have the same form, the proof that the fields given by the equations (61) and (62) are solutions of the second set of Einstein-Maxwell equations (55) and (53) is the same as in the case of the magnetic field.

One can easily verify that the electromagnetic duality implies the existence of a relationship between the scalar fields ϕ and θ that must satisfy the following equations simultaneously [11]

$$f(\phi, \bar{\phi}) \varepsilon^{ijk} D_j \phi D_k \bar{\phi} = \frac{g(\theta, \bar{\theta})}{N} \left[\left(\mathcal{L}_{\mathbf{m}} \bar{\theta} \right) D^i \theta - \left(\mathcal{L}_{\mathbf{m}} \theta \right) D^i \bar{\theta} \right], \quad (63)$$

$$g(\theta, \bar{\theta}) \varepsilon^{ijk} D_j \bar{\theta} D_k \theta = \frac{f(\phi, \bar{\phi})}{N} \left[\left(\mathcal{L}_{\mathbf{m}} \bar{\phi} \right) D^i \phi - \left(\mathcal{L}_{\mathbf{m}} \phi \right) D^i \bar{\phi} \right]. \quad (64)$$

These equations form a set of non-linear local constraints on the functions f and g and they must be satisfied at all times. There are no other constraints on f and g which shows that there are actually families of field line solutions rather than isolated solutions. One of these solutions is particularly interesting because it is a generalization of the Rañada field.

4.2. Local Generalization of Rañada's Solution

As discussed in the chapter [5] from this volume, the first field line solution of Maxwell's equations was given by Rañada in [2, 3]. The Rañada field is given in terms of two scalar functions f and g that play the same role as the scalar fields that have shown up in the previous subsection. The main difference between the two cases is that the form of f and g is fixed in the Rañada field to the following functions

$$f = \frac{1}{2\pi i} \frac{1}{(1 + |\phi|^2)^2}, \quad (65)$$

$$g = \frac{1}{2\pi i} \frac{1}{(1 + |\theta|^2)^2}. \quad (66)$$

It is interesting to see what is the gravitating electromagnetic field that is obtained by using the functions from the equations (65) and (66) in the vector fields $E^i(t,\mathbf{x})$ and $B^i(t,\mathbf{x})$ discussed above. By plugging these equations into the electric and magnetic field line solutions obtained in the previous subsection, and after some algebraic manipulation of the vector fields, we obtain the following result

$$B^i(t,\mathbf{x}) = \varepsilon^{ijk} D_j\alpha_1(t,\mathbf{x}) D_k\alpha_2(t,\mathbf{x}), \tag{67}$$

$$E^i(t,\mathbf{x}) = \varepsilon^{ijk} D_j\beta_1(t,\mathbf{x}) D_k\beta_2(t,\mathbf{x}), \tag{68}$$

where α_1, α_2, β_1 and β_2 are real scalar fields related to the complex fields ϕ and θ by the following relations

$$\alpha_1 = \frac{1}{1+|\phi|^2}, \qquad \alpha_2 = \frac{\Phi}{2\pi}, \qquad \phi = |\phi| e^{i\Phi}, \tag{69}$$

$$\beta_1 = \frac{1}{1+|\theta|^2}, \qquad \beta_2 = \frac{\Theta}{2\pi}, \qquad \theta = |\theta| e^{i\Theta}. \tag{70}$$

The fields given by the equations (67) and (68) represent the local generalization of the Rañada solution to the hyperbolic space-time M. It is important to note that the boundary conditions on ϕ and θ that generate the electromagnetic knots in the Minkowski space-time cannot be imposed automatically on the fields from the equations (69) and (70) since the limit $\mathbf{x} \rightarrow \infty$ is not well defined in the neighbourhood U_p in general.

We conclude this section by observing that the gravitating electromagnetic field can be analysed in close analogy with the electromagnetic field in flat space-time in the $1+3$-formalism. In this way, we can give a simple interpretation to the physical fields in terms of local geometrical and topological quantities. Also, the flat space-time limit can be easily obtained by choosing the Gauss normal coordinates systems with $N = 1, \boldsymbol{\beta} = 0$. The fields $E^i(t,\mathbf{x})$ and $B^i(t,\mathbf{x})$ take the form of Rañada's solutions in this frame. This fact leads to the conclusion that the gravitating field lines solutions represent a natural local generalization of Rañada's solutions. As mentioned above, it is possible to obtain global solutions if the topology of M does not impose any obstruction to the existence of global differential 2-forms.

5. THE KOPIŃSKI-NATÁRIO FIELD

In this section, we will present a particular field line solution that can be extended globally in the *Einstein universe* [12].

Let us recall the line element of the cosmological FRW metric [15] that is given by the following relation

$$ds^2 = -dt^2 + a(t)^2 \left[\frac{dr^2}{1 - kr^2} + r^2 \left(d\theta^2 + \sin^2(\theta) d\phi^2 \right) \right] . \tag{71}$$

The parameter a and the variable r can be rescaled such that k take the integer values $-1, 0$ or $+1$. If $k = 0$, the universe is flat, if $k = +1$, $r = \sin(\chi)$ the universe is closed, and if $k = -1$, $r = \sin\psi$ the universe is open. In what follows, we are interested in the foliated space-time with leaves Σ isomorphic to S^3. This correspond to a close universe. Matter content can be added to the model such that the homogeneity and the isotropy are preserved. Then the matter content is characterized by the following energy-momentum tensor

$$T_{\mu\nu} = (\rho + P)\, u_\mu u_\nu + P g_{\mu\nu} , \tag{72}$$

where u^μ is an unitary time-like vector, ρ is the density of mass and P is the density of pressure. Although $T_{\mu\nu}$ is the energy-momentum tensor of a perfect fluid, many matter models can be put into this form, including the electromagnetic field. In the co-moving frame, one can choose $u^\mu = (1, 0, 0, 0)$ and the energy-momentum tensor becomes diagonal $T^\mu{}_\nu = \mathrm{diag}(-\rho, P, P, P)$. Recall the Einstein's equations with a cosmological constant Λ

$$G_{\mu\nu} + \Lambda g_{\mu\nu} = 8\pi T_{\mu\nu} . \tag{73}$$

If the FRW metric from the equation (71) and the energy-momentum tensor from the equation (72) are substituted into the equation (73), the following set of equations is obtained

$$3\frac{\dot{a}^2 + k}{a^2} = 8\pi\rho + \Lambda , \tag{74}$$

$$\frac{2a\ddot{a} + \dot{a}^2 + k}{a^2} = 8\pi P + \Lambda , \tag{75}$$

$$\frac{\ddot{a}}{a^2} = -\frac{4\pi}{3}(\rho + 3P) + \frac{\Lambda}{3} . \tag{76}$$

We note that the equations (74)-(76) are not all independent of each other. These are the fundamental equations of cosmology and they describe a large variety of cosmological models that are homogeneous and isotropic. In particular, the *Einstein universe* is characterized by the following constraints

$$\dot{a} = \ddot{a} = 0\,, \qquad P = 0\,. \tag{77}$$

By using the equation (77) together with the cosmological equations (74)-(76), the following constitutive equations of the Einstein universe are obtained

$$\frac{3k}{a^2} = \Lambda + 8\pi\rho\,, \qquad \frac{k}{a^2} = \Lambda\,, \qquad k = 4\pi a^2 \rho\,. \tag{78}$$

The Einstein universe is a homogeneous model in space and time, but it is unstable to perturbations $a \to a + \varepsilon$, where $\varepsilon << 1$. Also, for a baryonic matter content, it requires that $k = +1$, so the universe is closed. One can determine its parameters up to a constant C and the result is given by the following relations

$$a = \frac{3C}{2}\,, \qquad \Lambda = \frac{4}{9C^2}\,. \tag{79}$$

The Einstein universe is not physical since the experimental data favours an expanding universe. Nevertheless, it is still an interesting model to be explored from both mathematical and physical point of view due to its highly homogeneous structure.

It is convenient to represent the Einstein universe in a coordinate system that displays the spherical symmetries of leafs. Since the space-time is foliated and has the topology $M \simeq \mathbb{R} \times S^3$, we can introduce an adapted coordinate system (t, x^i) at any point $p \in M$ and associate to it the canonical basis $(\mathbf{e}_0 = \partial_t, \mathbf{e}_i = \partial_i)$ of the tangent space $\mathcal{T}_p(M)$. Then according to Weyl's postulate, $\mathbf{e}_0$ is orthogonal to the leaf $\Sigma_t \simeq S^3$

$$g(\mathbf{e}_0\,, \mathbf{e}_i) = 0\,. \tag{80}$$

We use the fact that the leaf space can be represent as a group $S^3 \simeq SU(2)$. Then one can choose the basis $(\mathbf{e}_i) \in \mathcal{T}_p(S^3)$ such that the $su(2)$ algebra is satisfied by its elements

$$\mathbf{e}_0 = \partial_t \in \mathcal{T}_p(\mathbb{R})\,, \tag{81}$$

$$[\mathbf{e}_i\,, \mathbf{e}_j] = 2\varepsilon_{ijk}\mathbf{e}_k\,, \qquad \mathbf{e}_i \in \mathcal{T}_p(S^3)\,. \tag{82}$$

Note that the authors of [12] used the notation $\mathbf{e}_\mu = X_\mu$ so let us adopt it in what follows.

The dual basis to $(\mathbf{e}_\mu)$, denoted by $(\boldsymbol{\theta}^\mu)$, has the same decomposition along the directions of the foliated space-time manifold M. The elements of the dual basis also obey the $su(2)$ algebra since they satisfy the following equation

$$d\boldsymbol{\theta}^i = -\varepsilon_{ijk}\, d\boldsymbol{\theta}^j \wedge d\boldsymbol{\theta}^k \,. \tag{83}$$

One important feature of the Einstein universe is that it is locally conformal equivalent to the Minkowski space-time. By using this equivalence, the local Einstein-Maxwell equations take the same form as the Maxwell equations in some region of the flat space-time. This is possible since the Maxwell equations are invariant under the conformal transformations [12]. As discussed in the previous sections, they have the following form in the absence of sources

$$dF = 0\,, \qquad d \star F = 0\,. \tag{84}$$

The decomposition of the electromagnetic 2-form in to electric and magnetic components is standard and it is given by the following relation

$$F = E^i\, \boldsymbol{\theta}^i \wedge \boldsymbol{\theta}^0 + \frac{1}{2} B^i \varepsilon_{ijk}\, \boldsymbol{\theta}^j \wedge \boldsymbol{\theta}^k \,. \tag{85}$$

By plugging the equation (85) into the equations (84), one can easily obtain the following set of equations

$$X_i(E^i) = X_i(B^i) = 0\,, \tag{86}$$

$$\dot{B}^i - 2E^i + \varepsilon_{ijk} X_j(E^k) = 0\,, \tag{87}$$

$$\dot{E}^i + 2B^i - \varepsilon_{ijk} X_j(B^k) = 0\,. \tag{88}$$

The Kopiński-Natário field is obtain by making the ansatz that the components of the electric and magnetic fields depend only on the time as measured in a stationary frame [12]. This assumption is valid in the Einstein universe which is a stationary closed model. As such, the variations along the directions of the $SU(2)$ manifold vanish and the equations (87) and (88) take the following form

$$\dot{B}^i - 2E^i = 0\,, \tag{89}$$

$$\dot{E}^i + 2B^i = 0\,. \tag{90}$$

The simplest solution of the equations (89) and (90) obtained in [12] has the following form

$$E(t) = E_0 \cos(2t) X_1 \,, \tag{91}$$

$$B(t) = B_0 \cos(2t) X_1 \,. \tag{92}$$

However, one can see that the equations (89) and (90) admit many other solutions generated by the second degree differential equation

$$\ddot{f} \pm 4f = 0 \,. \tag{93}$$

One important remark is that the components of the Kopiński-Natário field are not orthogonal to each other as they share the same direction of $SU(3)$. Also, their field lines are unstable under small perturbations.

As the authors noted in their paper [12], the Einstein universe and the Minkowski space-time are conformally equivalent to each other in some open regions, which allows one to conformally map the objects defined in the Einstein universe into similar objects that live in the Minkowski space-time. The conformal mapping is defined by the following transformation of the metric tensor

$$g_{\mu\nu} = \Omega \, \eta_{\mu\nu} \,, \tag{94}$$

where $\eta_{\mu\nu}$ is the metric of Minkowski space-time and Ω is the conformal factor. In particular, if one interprets the energy of the field line solution in terms of the energy defined in the flat space-time, the following relation is found [12]

$$\mathcal{E}(t) = \frac{E_0^2}{12} \left[9\pi \sin(t) + \pi \sin(3t) + 12\pi^2 \cos(t) - 12\pi t \cos(t) \right] \,, \tag{95}$$

The energy $\mathcal{E}(t)$ shows that the Kopiński-Natário solution describes a radiation field.

In the reference [12], similar considerations were made for a general FRW closed universe described by the line element from the equation (71). In that case, the vector fields corresponding to the metric are given by the following relations

$$\bar{\mathbf{X}}_0 = \partial_t \,, \qquad \bar{\mathbf{X}}_i = a^{-1}(t) \partial_i \,. \tag{96}$$

The dual basis associated to them is composed by the following tetrad fields

$$\bar{\theta}^0 = dt \,, \qquad \bar{\theta}^i = a(t) \bar{\theta}^i \,, \tag{97}$$

The electromagnetic 2-form field can be decomposed with respect to this basis and the following relation is obtained

$$F = -E^1 \bar{\theta}^0 \wedge \bar{\theta}^1 + B^1 \bar{\theta}^2 \wedge \bar{\theta}^3 . \tag{98}$$

Then the Einstein-Maxwell equations in the vacuum take the following form

$$\frac{d}{dt}\left[a^2(t)B^1\right] = 2aE^1 , \tag{99}$$

$$\frac{d}{dt}\left[a^2(t)E^1\right] = -2aB^1 , \tag{100}$$

The solution of the equations (101) and (102) can be easily obtained by integration. The result is given by the following relations

$$E^1(t) = \frac{E_0}{a^2(t)} \cos\left[\int dt\, a^{-1}(t)\right] , \tag{101}$$

$$B^1(t) = \frac{E_0}{a^2(t)} \sin\left[\int dt\, a^{-1}(t)\right] . \tag{102}$$

The energy of this solution goes as $\sim E_0^2 a^{-4}(t)$. The Einstein universe represents a particular case of the above equations for which a is constant. For a more complete discussion of the properties of this solution we refer the reader to the original work [12].

APPENDIX

In this appendix, we collect some definitions and properties of the Lie derivative, the interior derivative and the extrinsic curvature that have been used in the text. This is a standard material and can be found in any classic text on differential geometry. For applications in physics, see e. g. [17].

In all the definitions reviewed here, we will consider a differentiable manifold M of dimension $\dim(M) = n$.

Definition 5.1. *If $T \in \mathcal{T}^p_q(M)$ is a tensor field of rank (p, q) and $X \in \mathcal{X}(M)$ is a differentiable vector field, then the* Lie derivative *of T along X is defined as follows*

$$(\mathcal{L}_Y T)_p = \left.\frac{d}{dt}\right|_{t=0} \left[(\mu_{-t})_* T_{\varphi_t(p)}\right] = \left.\frac{d}{dt}\right|_{t=0} \left[(\mu_t)^* T_p\right] . \tag{103}$$

Here, $\mu : I \times M \to M$, $I \subset \mathbb{R}$ is the one-parameter semigroup of diffeomorphisms on M generated by the flow of X with the action

$$x \to \mu_t(x) = \mu(t,x)\,,\ \forall x \in M\,, \qquad X(x) = \left.\frac{d\mu(t,x)}{dt}\right|_{t=0}. \tag{104}$$

The Lie derivative obeys the following axioms

$$\mathcal{L}_X f = X(f)\,, \tag{105}$$
$$[\mathcal{L}_X\,,\,d] = 0\,, \tag{106}$$
$$\mathcal{L}_X(T \otimes S) = (\mathcal{L}_X T) \otimes S + T \otimes (\mathcal{L}_X S)\,, \tag{107}$$
$$\mathcal{L}_X\left(T(X_1,\ldots,X_n\right) = \left(\mathcal{L}_X T\right)\left(X_1,\ldots,X_n\right) + T\left(\mathcal{L}_X X_1,\ldots,X_n\right) + \cdots + T\left(X_1,\ldots,\mathcal{L}_X X_n\right)\,, \tag{108}$$

where T and S are arbitrary tensor fields and d is the exterior derivative. In the above relations, the objects considered there have natural properties. It is easy to show that the following relations hold

$$\mathcal{L}_X Y(f) = X(Y(f)) - Y(X(f)) = [X,Y](f)\,, \tag{109}$$
$$\mathcal{L}_X \omega = \iota_X d\omega + d\iota_X \omega\,. \tag{110}$$

In the equation (110), we have denoted by ω a differential form on M and by ι the interior product to be defined below.

Definition 5.2. *Let $X \in \mathcal{X}(M)$ be a smooth vector field on M and $\omega \in \Omega^k(M)$ a k-form. Then the* interior product *of X and ω is a $(k-1)$-form $\iota_X\omega$ defined by the following property*

$$\iota_X \omega(X_1,\ldots,X_{k-1}) = \omega(X, X_1,\ldots X_{k-1})\,. \tag{111}$$

If $k = 0$ we take by definition $\iota_X\omega = 0$.

It is easy to show that the following equalities are true

$$\iota_X(\omega \wedge \gamma) = \iota_X\omega \wedge \gamma + (-1)^k \omega \wedge \iota_X \gamma\,,\ \forall \omega \in \Omega^k(M)\,, \forall \gamma \in \Omega^q(M)\,, \tag{112}$$
$$\iota_{[X,Y]} = [\mathcal{L}_X, \iota_Y]\,, \qquad \forall X, Y \in \mathcal{X}(M)\,, \tag{113}$$
$$\iota_X \iota_Y \omega = -\iota_Y \iota_X \omega\,,\ \forall X, Y \in \mathcal{X}(M)\,, \forall \omega \in \Omega^k(M)\,. \tag{114}$$

Let us we review the *extrinsic curvature* of an embedded surface in three dimensions. The generalization to four-dimensions is straightforward. Consider an oriented surface $\Sigma \in \mathbb{R}^3$ on which the orientation is defined by the unit vector field $\mathbf{n}$. Then one can define the *Gauss map* as follows

$$\boldsymbol{\nu} : \Sigma \to S^2 , \qquad \boldsymbol{\nu}(x) = \mathbf{n}(x) . \tag{115}$$

Since $\boldsymbol{\nu}$ is smooth, it induces the following map

$$D_x\boldsymbol{\nu} : \mathcal{T}_x(\Sigma) \to \mathcal{T}_{\boldsymbol{\nu}(x)}(S^2) . \tag{116}$$

We know that $\mathcal{T}_{\boldsymbol{\nu}(x)}(S^2) \simeq \mathcal{T}_x(\Sigma)$. Then the derivative $D_x\boldsymbol{\nu}$ defines the *Weingarten map* as follows

$$\mathcal{W}_x = -D_x\boldsymbol{\nu} : \mathcal{T}_x(\Sigma) \to \mathcal{T}_x(\Sigma) . \tag{117}$$

The symmetric bilinear two-form curvature map is obtained from $\mathcal{W}_x$ as follows

$$K_x(X, Y) = \langle \mathcal{W}_x X, Y \rangle \tag{118}$$

for any $X, Y \in \mathcal{T}_x(\Sigma)$.

REFERENCES

[1] Trautman, A. (1977). "Solutions of the Maxwell and Yang-Mills Equations Associated with Hopf Fibrings," *Int. J. Theor. Phys.* 16, 561.

[2] Rañada, A. F. (1989). "A Topological Theory of the Electromagnetic Field," *Lett. Math. Phys.* 18, 97.

[3] Rañada, A. F. (1990). "Knotted solutions of the Maxwell equations in vacuum," *J. Phys. A* 23, L815.

[4] Arrayás, M., Bouwmeester, D. and Trueba, J. L. (2017). "Knots in electromagnetism," *Phys. Rept.* 667.

[5] Vancea, I. V. (2019). "Knots and the Maxwell Equations," chapter in this volume.

[6] Dalhuisen, J. W. and Bouwmeester, D. (2012). "Twistors and electromagnetic knots," *J. Phys. A* 45, 135201.

[7] Swearngin, J., Thompson, A., Wickes, A., Dalhuisen, J. W. and Bouwmeester, D. (2013). "Gravitational Hopfions," arXiv:1302.1431 [gr-qc].

[8] Thompson, A., Swearngin, J. and Bouwmeester, D. (2014). 'Linked and Knotted Gravitational Radiation," *J. Phys. A* 47, 355205.

[9] Thompson, A., Wickes, A., Swearngin, J. and Bouwmeester, D. (2015). "Classification of Electromagnetic and Gravitational Hopfions by Algebraic Type," *J. Phys. A* 48, 205202.

[10] Alves, D. W. F. and Nastase, H. (2018). "Hopfion solutions in gravity and a null fluid/gravity conjecture," arXiv:1812.08630 [hep-th].

[11] Vancea, I. V. (2017). "On the existence of the field line solutions of the Einstein - Maxwell equations," *Int. J. Geom. Meth. Mod. Phys.* 15, 1850054.

[12] Kopiński, J. and Natário, J. (2017). "On a remarkable electromagnetic field in the Einstein Universe," *Gen. Rel. Grav.* 49, 81.

[13] Costa e Silva, W., Goulart, E. and Ottoni, J. E. (2018). "On spacetime foliations and electromagnetic knots," arXiv:1809.09259 [math-ph].

[14] Gourgoulhon, É. (2012). "3+1 Formalism in General Relativity: Bases of Numerical Relativity," Lecture Notes in Physics, Volume 846, Springer-Verlag.

[15] Weinberg, S. (1972). "Gravitation and Cosmology : Principles and Applications of the General Theory of Relativity," John Wiley and Sons.

[16] Jackson, J. D. (1962). "Classical Electrodynamics," John Wiley and Sons.

[17] Frankel, T. (1997). "The geometry of physics: An introduction," Cambridge University Press.

[18] Deser, S. and Teitelboim, C. (1976). "Duality Transformations of Abelian and Nonabelian Gauge Fields," *Phys. Rev. D* 13, 1592.

In: An Essential Guide to Maxwell's Equations ISBN: 978-1-53616-680-4
Editor: Casey Erickson

Chapter 3

EXISTENCE OF A WEAK SOLUTION IN AN EVOLUTIONARY MAXWELL-STOKES TYPE PROBLEM AND THE ASYMPTOTIC BEHAVIOR OF THE SOLUTION

Junichi Aramaki*
Division of Science, Tokyo Denki University, Tokyo, Japan

Abstract

We consider the existence of a weak solution to a class of an evolutionary Maxwell-Stokes type problem containing a p-curlcurl system in a multi-connected domain. Moreover, we show that the solution converges to a solution of the stationary Maxwell-Stokes type problem as the time tending to the infinity.

PACS: 05.45-a, 52.35.Mw, 96.50.Fm

Keywords: Maxwell-Stokes system, weak solution, p-curlcurl operator, asymptotic behavior, multi-connected domain

AMS Subject Classification: 35K55, 35K60, 35K65, 35K90

*Corresponding Author's E-mail: aramaki@hctv.ne.jp.

1. INTRODUCTION

Generalized Maxwell's equations in electromagnetic field are written by

$$\begin{cases} \varepsilon \boldsymbol{E}_t + \sigma \boldsymbol{j} = \operatorname{curl} \boldsymbol{H}, \\ \mu \boldsymbol{H}_t + \operatorname{curl} \boldsymbol{E} = \boldsymbol{F}, \\ \varepsilon \operatorname{div} \boldsymbol{E} = q, \\ \operatorname{div} \boldsymbol{H} = 0 \end{cases} \tag{1.1}$$

in $\Omega_T := \Omega \times (0,T)$, where Ω is a bounded domain in $\mathbb{R}^3$ with a boundary Γ, $\boldsymbol{E}$ and $\boldsymbol{H}$ denote the electric and the magnetic fields, respectively, ε is the permittivity of the electric field, μ is the permeability of the magnetic field, σ is the electric conductivity of the material, $\boldsymbol{j}$ is the total current density and q is the density of electric charge. Since the displace current $\varepsilon \boldsymbol{E}_t$ is small in comparison with eddy currents, we neglect the term. We use the nonlinear extension of Ohm's law $|\boldsymbol{j}|^{p-2}\boldsymbol{j} = \sigma \boldsymbol{E}$. Then $\boldsymbol{H}$ satisfies the following equations containing p-curlcurl equation

$$\begin{cases} \mu \boldsymbol{H}_t + \operatorname{curl}\left[\frac{1}{\sigma}|\operatorname{curl} \boldsymbol{H}|^{p-2}\operatorname{curl} \boldsymbol{H}\right] = \boldsymbol{F}, \\ \operatorname{div} \boldsymbol{H} = 0 \end{cases} \tag{1.2}$$

in Ω_T. We impose the natural boundary condition

$$\boldsymbol{H} \cdot \boldsymbol{n} = 0 \text{ and } \sigma^{p-1} \boldsymbol{E} \times \boldsymbol{n} = \boldsymbol{h} \times \boldsymbol{n} \text{ on } \Gamma_T := \Gamma \times (0,T), \tag{1.3}$$

where $\boldsymbol{n}$ denotes the outward normal unit vector field to Γ and we also impose the initial condition

$$\boldsymbol{H}(0) = \boldsymbol{H}_0 \text{ on } \Omega. \tag{1.4}$$

Putting $\nu = 1/\sigma$, we consider the following system.

$$\begin{cases} \mu \boldsymbol{H}_t + \operatorname{curl}[\nu|\operatorname{curl} \boldsymbol{H}|^{p-2}\operatorname{curl} \boldsymbol{H}] = \boldsymbol{F} & \text{in } \Omega_T, \\ \operatorname{div} \boldsymbol{H} = 0 & \text{in } \Omega_T, \\ \boldsymbol{H} \cdot \boldsymbol{n} = 0, & \text{on } \Gamma_T, \\ \nu|\operatorname{curl} \boldsymbol{H}|^{p-2}\operatorname{curl} \boldsymbol{H} \times \boldsymbol{n} = \boldsymbol{h} \times \boldsymbol{n} & \text{on } \Gamma_T, \\ \boldsymbol{H}(0) = \boldsymbol{H}_0 & \text{in } \Omega. \end{cases} \tag{1.5}$$

As a necessary condition for the existence of a solution to this problem, the external field $\boldsymbol{F}$ must satisfy $\operatorname{div} \boldsymbol{F} = 0$ in Ω_T. Moreover, since

$$\boldsymbol{n} \cdot \operatorname{curl}[\nu|\operatorname{curl} \boldsymbol{H}|^{p-2}\operatorname{curl} \boldsymbol{H}] = \operatorname{div}_\Gamma(\nu|\operatorname{curl} \boldsymbol{H}|^{p-2}\operatorname{curl} \boldsymbol{H} \times \boldsymbol{n}) \text{ in } \Gamma_T,$$

where div_Γ denotes the surface divergence (cf. Mitreau et al. [11]), an another necessary condition for the existence of a solution to this problem is

$$\boldsymbol{F}\cdot\boldsymbol{n} = \mathrm{div}_\Gamma(\boldsymbol{h}\times\boldsymbol{n}) \text{ on } \Gamma_T.$$

Yin et al. [14] obtained the existence theorem of a weak solution of (1.5) in the case where $p > 2$, Ω is a bounded simply connected domain without holes, and under the boundary condition $\boldsymbol{H}\times\boldsymbol{n} = \boldsymbol{0}$ instead of $\boldsymbol{H}\cdot\boldsymbol{n} = 0$. Miranda et al [10] considered the problem (1.5) under more general setup in the case where Ω is simply connected. They obtained a "weak" solution. However, their "weak" solution is not the solution of (1.5) in the distribution sense, so the "weak" solution in not the weak solution strictly speaking.

In this chapter, we consider a more general system than (1.5) in the case where Ω is a multi-connected domain.

$$\begin{cases} \partial_t\boldsymbol{u} + \mathrm{curl}\,[S_s(x,t,|\mathrm{curl}\,\boldsymbol{u}|^2)\mathrm{curl}\,\boldsymbol{u}] = \boldsymbol{F} & \text{in } \Omega_T, \\ \mathrm{div}\,\boldsymbol{u} = 0 & \text{in } \Omega_T, \\ \boldsymbol{u}\cdot\boldsymbol{n} = 0, & \text{on } \Gamma_T, \\ S_s(x,t,|\mathrm{curl}\,\boldsymbol{u}|^2)\mathrm{curl}\,\boldsymbol{u}\times\boldsymbol{n} = \boldsymbol{h}\times\boldsymbol{n} & \text{on } \Gamma_T, \\ \langle\boldsymbol{u}\cdot\boldsymbol{n},1\rangle_{\Sigma_j} = 0 & j = 1,\dots,J, \\ \boldsymbol{u}(0) = \boldsymbol{u}_0 & \text{in } \Omega, \end{cases} \tag{1.6}$$

where the function $S(x,t,s)$ satisfies some structure conditions, and Σ_j $(j = 1,\dots,J)$ are cuts of Ω such that $\Omega\setminus(\cup_{j=1}^J\Sigma_j)$ is simply connected and $\langle\boldsymbol{u}\cdot\boldsymbol{n},1\rangle_{\Sigma_j}$ denotes the duality bracket. More precisely, these are defined in section 2. Under the hypothesis $\mathrm{div}\,\boldsymbol{F} = 0$ in Ω_T and $\boldsymbol{F}\cdot\boldsymbol{n} = \mathrm{div}_\Gamma(\boldsymbol{h}\times\boldsymbol{n})$ on Γ_T, we show the existence of a weak solution of (1.6). Our weak solution satisfies (1.6) in the distribution sense. To show this, we must extend the space of test functions in the weak formulation of (1.6).

In the case where the condition $\mathrm{div}\,\boldsymbol{F} = 0$ in Ω_T is not satisfied, it is natural to consider the following Maxwell-Stokes type problem: to find $(\boldsymbol{u},\pi)$ in an appropriate space such that

$$\begin{cases} \partial_t\boldsymbol{u} + \mathrm{curl}\,[S_s(x,t,|\mathrm{curl}\,\boldsymbol{u}|^2)\mathrm{curl}\,\boldsymbol{u}] + \nabla\pi = \boldsymbol{f} & \text{in } \Omega_T, \\ \mathrm{div}\,\boldsymbol{u} = 0 & \text{in } \Omega_T, \\ \boldsymbol{u}\cdot\boldsymbol{n} = 0, & \text{on } \Gamma_T, \\ S_s(x,t,|\mathrm{curl}\,\boldsymbol{u}|^2)\mathrm{curl}\,\boldsymbol{u}\times\boldsymbol{n} = \boldsymbol{h}\times\boldsymbol{n} & \text{on } \Gamma_T, \\ \langle\boldsymbol{u}\cdot\boldsymbol{n},1\rangle_{\Sigma_j} = 0 & j = 1,\dots,J, \\ \boldsymbol{u}(0) = \boldsymbol{u}_0 & \text{in } \Omega. \end{cases} \tag{1.7}$$

We derive the existence of a unique weak solution $(\boldsymbol{u},\pi)$ of (1.7) using the result on the existence of a weak solution of (1.6).

It is also interesting to consider the asymptotic behavior of a weak solution $(\boldsymbol{u},\pi) = (\boldsymbol{u}(t),\pi(t))$ as $t \to \infty$. We show that $(\boldsymbol{u}(t),\pi(t))$ converges to a weak solution of the stationary version of (1.7).

The chapter is organized as follows. In section 2, we give some preliminaries on the shape of the domain Ω, some spaces of functions and the structure conditions of a Carathéodory function $S(x,t,s)$. In section 3, we consider the problem (1.6). Section 4 is devoted to the existence of a unique solution of (1.7). Finally, in section 5, we consider the asymptotic behavior of the solution obtained in section 4 as the time t tends to the infinity.

2. Preliminaries

In this section, we state some preliminaries. Let Ω be a bounded domain in $\mathbb{R}^3$ with a $C^{1,1}$ boundary Γ, and $1 < p < \infty$. We denote the conjugate exponent of p by p', i.e., $(1/p) + (1/p') = 1$. From now on we use $L^p(\Omega)$, $W^{m,p}(\Omega)$, $H^m(\Omega)$, $W^{s,p}(\Gamma)$ and $H^s(\Gamma)$ for the standard L^p and Sobolev spaces of functions defined in Ω and Γ, respectively. For any Banach space B, we denote $B \times B \times B$ by boldface character $\boldsymbol{B}$. Hereafter, we use this character to denote vectors and vector-valued functions, and we denote the standard Euclidean inner product of vectors $\boldsymbol{a}$ and $\boldsymbol{b}$ by $\boldsymbol{a} \cdot \boldsymbol{b}$. For the dual space $\boldsymbol{B}'$ of $\boldsymbol{B}$, we denote $\langle \cdot, \cdot \rangle_{\boldsymbol{B}',\boldsymbol{B}}$ for the duality bracket.

Since we allow Ω to be a multi-connected domain in $\mathbb{R}^3$, we assume that Ω satisfies the following conditions as in Amrouche and Seloula [1] (cf. also see Amrouche and Seloula [2], Dautray and Lions [6] and Girault and Raviart [7]). Ω is locally situated on one side of Γ and satisfies the following (O1) and (O2).

(O1) Γ has a finite number of connected components $\Gamma_0, \Gamma_1, \ldots, \Gamma_I$ with Γ_0 denoting the boundary of the infinite connected component of $\mathbb{R}^3 \setminus \overline{\Omega}$.

(O2) There exist J connected open surfaces Σ_j, $(j = 1, \ldots, J)$, called cuts, contained in Ω such that

   (a) Σ_j is an open subset of a smooth manifold $\mathcal{M}_j$.

   (b) $\partial\Sigma_j \subset \Gamma$ $(j = 1, \ldots, J)$, where $\partial\Sigma_j$ denotes the boundary of Σ_j, and Σ_j is non-tangential to Γ.

(c) $\overline{\Sigma_j} \cap \overline{\Sigma_k} = \emptyset \, (j \neq k)$.

(d) The open set $\Omega^\circ = \Omega \setminus (\cup_{j=1}^J \Sigma_j)$ is simply connected and pseudo $C^{1,1}$ class.

The number J is called the first Betti number and I the second Betti number. We say that Ω is simply connected if $J = 0$ and Ω has no holes if $I = 0$. If we define

$$\mathbb{K}_T^p(\Omega) = \{\boldsymbol{v} \in \boldsymbol{L}^p(\Omega); \operatorname{curl} \boldsymbol{v} = \boldsymbol{0}, \operatorname{div} \boldsymbol{v} = 0 \text{ in } \Omega, \boldsymbol{v} \cdot \boldsymbol{n} = 0 \text{ on } \Gamma\}$$

and

$$\mathbb{K}_N^p(\Omega) = \{\boldsymbol{v} \in \boldsymbol{L}^p(\Omega); \operatorname{curl} \boldsymbol{v} = \boldsymbol{0}, \operatorname{div} \boldsymbol{v} = 0 \text{ in } \Omega, \boldsymbol{v} \times \boldsymbol{n} = \boldsymbol{0} \text{ on } \Gamma\},$$

then it is well known that $\dim \mathbb{K}_T^p(\Omega) = J$ and $\dim \mathbb{K}_N^p(\Omega) = I$. In the later, we need a basis of $\mathbb{K}_T^p(\Omega)$ as following. Let $q_j^T \in W^{2,p}(\Omega^\circ)$ be a unique solution of the problem

$$\begin{cases} -\Delta q_j^T = 0 & \text{in } \Omega^\circ, \\ \partial_{\boldsymbol{n}} q_j^T = 0 & \text{on } \Gamma, \\ \left[q_j^T\right]_{\Sigma_k} = \text{const., and } \left[\partial_{\boldsymbol{n}} q_j^T\right]_{\Sigma_k} = 0 & \text{for } k = 1, \ldots, J, \\ \langle \partial_{\boldsymbol{n}} q_j^T, 1 \rangle_{\Sigma_k} = \delta_{jk} & \text{for } k = 1, \ldots, J, \end{cases}$$

where $\left[q_j^T\right]_{\Sigma_k}$ denotes the jump across Σ_k and

$$\langle \cdot, \cdot \rangle_{\Sigma_k} = \langle \cdot, \cdot \rangle_{W^{-1/p,p}(\Sigma_k), W^{1/p,p'}(\Sigma_k)}.$$

Since $\nabla q_j^T \in L^p(\Omega^\circ)$, it can be extended to a function in $\boldsymbol{L}^p(\Omega)$ and we denote it by $\widetilde{\nabla} q_j^T$. Then we can see that $\{\widetilde{\nabla} q_j^T\}_{j=1}^J$ is a basis of $\mathbb{K}_T^p(\Omega)$ (cf. [1, Corollary 4.1]).

We introduce some spaces of vector functions. Define a space

$$\mathbb{X}^p(\Omega) = \{\boldsymbol{v} \in \boldsymbol{L}^p(\Omega); \operatorname{curl} \boldsymbol{v} \in \boldsymbol{L}^p(\Omega), \operatorname{div} \boldsymbol{v} \in L^p(\Omega)\}$$

with the norm

$$\|\boldsymbol{v}\|_{\mathbb{X}^p(\Omega)} = \|\boldsymbol{v}\|_{\boldsymbol{L}^p(\Omega)} + \|\operatorname{curl} \boldsymbol{v}\|_{\boldsymbol{L}^p(\Omega)} + \|\operatorname{div} \boldsymbol{v}\|_{L^p(\Omega)}.$$

Then $\mathbb{X}^p(\Omega)$ is a Banach space. We note that if $\boldsymbol{v} \in \boldsymbol{L}^p(\Omega)$ and $\operatorname{div} \boldsymbol{v} \in L^p(\Omega)$, then the normal trace $\boldsymbol{v} \cdot \boldsymbol{n} \in W^{-1/p,p}(\Gamma)$ is well defined and

$$\langle \boldsymbol{v} \cdot \boldsymbol{n}, \varphi \rangle_{W^{-1/p,p}(\Gamma), W^{1-1/p',p'}(\Gamma)} = \int_\Omega \boldsymbol{v} \cdot \nabla \varphi dx + \int_\Omega (\operatorname{div} \boldsymbol{v}) \varphi dx$$

for any $\varphi \in W^{1,p'}(\Omega)$ (cf. [1, p. 45]). Furthermore, define a closed subspace of $\mathbb{X}^p(\Omega)$ by

$$\mathbb{X}^p_T(\Omega) = \{v \in \mathbb{X}^p(\Omega); v \cdot n = 0 \text{ on } \Gamma\}.$$

Then $\mathbb{X}^p_T(\Omega) \hookrightarrow W^{1,p}(\Omega)$ and there exists a constant $C > 0$ depending only on p and Ω such that

$$\|v\|_{W^{1,p}(\Omega)} \leq C\|v\|_{\mathbb{X}^p(\Omega)} \text{ for all } v \in \mathbb{X}^p_T(\Omega).$$

Moreover, we define a space

$$\mathbb{V}^p_T(\Omega) = \{v \in \mathbb{X}^p_T(\Omega); \operatorname{div} v = 0 \text{ in } \Omega, \langle v \cdot n, 1\rangle_{\Sigma_j} = 0 \text{ for } j = 1, \ldots, J\}.$$

The following inequalities are used frequently (cf. [1]). If we define

$$\mathbb{X}^{1,p}(\Omega) = \{v \in \mathbb{X}^p(\Omega); v \cdot n \in W^{1-1/p,p}(\Gamma)\},$$

then we can see that $\mathbb{X}^{1,p}(\Omega) \hookrightarrow W^{1,p}(\Omega)$ and there exists a constant $C > 0$ depending only on p and Ω such that

$$\|v\|_{W^{1,p}(\Omega)} \leq C(\|v\|_{\mathbb{X}^p(\Omega)} + \|v \cdot n\|_{W^{1-1/p,p}(\Gamma)}). \tag{2.1}$$

Moreover, we can deduce the following (cf. [1, p. 40]). For any $v \in W^{1,p}(\Omega)$ with $v \cdot n = 0$ on Γ,

$$\|v\|_{L^p(\Omega)} + \|\nabla v\|_{L^p(\Omega)} \leq C\big(\|\operatorname{curl} v\|_{L^p(\Omega)} + \|\operatorname{div} v\|_{L^p(\Omega)} + \sum_{j=1}^{J} |\langle v \cdot n, 1\rangle_{\Sigma_j}|\big). \tag{2.2}$$

Thus we have the following.

Lemma 2.1. *$\mathbb{V}^p_T(\Omega)$ is a reflexive, separable Banach space with the norm*

$$\|v\|_{\mathbb{V}^p_T(\Omega)} := \|\operatorname{curl} v\|_{L^p(\Omega)}$$

which is equivalent to the $W^{1,p}(\Omega)$-norm.

We note that it follows from the Sobolev embedding theorem that there exists a constant $C > 0$ depending only on p and Ω such that for all $v \in \mathbb{V}^p_T(\Omega)$,

$$\|v\|_{L^p(\Omega)} + \|v\|_{L^p(\Gamma)} \leq C\|v\|_{\mathbb{V}^p_T(\Omega)},$$

where the second term of the left hand side denotes the norm of the trace of $\boldsymbol{v}$ on Γ. Then it follows from the Sobolev embedding theorem that if $6/5 \leq p < \infty$, then $\mathbb{X}_T^p(\Omega) \subset \boldsymbol{L}^2(\Omega)$. Moreover, if $6/5 \leq p < \infty$, define a Hilbert space

$$\boldsymbol{L}_\sigma^2(\Omega) = \text{ the closure of } \mathbb{V}_T^p(\Omega) \text{ in } \boldsymbol{L}^2(\Omega).$$

Then we have that $\mathbb{V}_T^p(\Omega) \hookrightarrow \mathbb{X}_T^p(\Omega)$ and so $\mathbb{X}_T^p(\Omega)' \hookrightarrow \mathbb{V}_T^p(\Omega)'$. $\mathbb{V}_T^p(\Omega)$ is dense in $\boldsymbol{L}_\sigma^2(\Omega)$ and $\mathbb{V}_T^p(\Omega) \hookrightarrow \boldsymbol{L}_\sigma^2(\Omega) \hookrightarrow \mathbb{V}_T^p(\Omega)'$.

Let $0 < T < \infty$, and let $\Omega_T = \Omega \times (0,T)$ and $\Gamma_T = \Gamma \times (0,T)$. Assume that $S(x,t,s)$ is a Carathéodoty function in $\Omega_T \times [0,\infty)$ satisfying the following conditions.

For a.e. $(x,t) \in \Omega_T$, $S(x,t,s) \in C^2((0,\infty)) \cap C([0,\infty))$ as a function of s, and there exist $1 < p < \infty$ and $0 < \lambda \leq \Lambda < \infty$ such that for a.e. $(x,t) \in \Omega_T$ and $s > 0$,

$$S(x,t,0) = 0 \text{ and } \lambda s^{(p-2)/2} \leq S_s(x,t,s) \leq \Lambda s^{(p-2)/2}. \tag{2.3a}$$

$$\lambda s^{(p-2)/2} \leq S_s(x,t,s) + 2sS_{ss}(x,t,s) \leq \Lambda s^{(p-2)/2}. \tag{2.3b}$$

$$\begin{cases} S_{ss}(x,t,s) < 0 & \text{if } 1 < p < 2, \\ S_{ss}(x,t,s) \geq 0 & \text{if } p \geq 2, \end{cases} \tag{2.3c}$$

where $S_s = \partial S/\partial s, S_{ss} = \partial^2 S/\partial s^2$. We note that (2.3a) implies that

$$\frac{2}{p}\lambda s^{p/2} \leq S(x,t,s) \leq \frac{2}{p}\Lambda s^{p/2} \text{ for a.e. } (x,t) \in \Omega_T \text{ and } s \in [0,\infty). \tag{2.4}$$

For a.e. $(x,t) \in \Omega_T$, it follows from (2.3b) that $G(s) = S(x,t,s^2)$ is a strictly convex function with respect to $s \in [0,\infty)$. Indeed, $G'(s) = 2sS_s(x,t,s^2)$ and so

$$G''(s) = 2(S_s(x,t,s^2) + 2s^2 G_{ss}(x,t,s^2)) > 0 \text{ for } s > 0.$$

Example 2.2. *If $S(x,t,s) = \nu(x,t)s^{p/2}$, where ν is a measurable function in Ω_T and satisfies $0 < \nu_* \leq \nu(x,t) \leq \nu^* < \infty$ for a.e. $(x,t) \in \Omega_T$ for some constants ν_* and ν^*, then it follows from elementary calculations that* (2.3a)-(2.3c) *hold.*

We give the following lemma with respect to monotonicity of S_s.

Lemma 2.3. *There exists a constant $c > 0$ such that for all $\boldsymbol{a}, \boldsymbol{b} \in \mathbb{R}^3$,*

$$(S_s(x,t,|\boldsymbol{a}|^2)\boldsymbol{a} - S_s(x,t,|\boldsymbol{b}|^2)\boldsymbol{b}) \cdot (\boldsymbol{a} - \boldsymbol{b}) \geq \begin{cases} c|\boldsymbol{a}-\boldsymbol{b}|^p & \text{if } p \geq 2, \\ c(|\boldsymbol{a}|+|\boldsymbol{b}|)^{p-2}|\boldsymbol{a}-\boldsymbol{b}|^2 & \text{if } 1 < p < 2. \end{cases}$$

In particular, S_s is strictly monotone, that is,

$$(S_s(x,t,|\boldsymbol{a}|^2)\boldsymbol{a} - S_s(x,t,|\boldsymbol{b}|^2)\boldsymbol{b}) \cdot (\boldsymbol{a} - \boldsymbol{b}) > 0 \text{ if } \boldsymbol{a} \neq \boldsymbol{b}.$$

For the proof, see Aramaki [4, Lemma 3.6].

3. Existence of a Weak Solution to an Evolution Problem

In this section, we consider the system (1.6).

Throughout this paper, we assume that $6/5 \le p < \infty$.

We give the formulation of a weak solution of (1.6).

Assume that $\boldsymbol{F} \in L^{p'}(0,T;\boldsymbol{L}^{p'}(\Omega))$ satisfies $\operatorname{div}\boldsymbol{F} = 0$ in Ω_T, $\boldsymbol{h}\times\boldsymbol{n} \in L^{p'}(0,T;\boldsymbol{W}^{-1/p',p'}(\Gamma))$ and $\boldsymbol{u}_0 \in \boldsymbol{L}^2_\sigma(\Omega)$.

Definition 3.1. *We say that* $\boldsymbol{u}$ *is a weak solution of* (1.6) *if*

$$\boldsymbol{u} \in L^p(0,T;\mathbb{V}^p_T(\Omega)) \cap C([0,T];\boldsymbol{L}_\sigma(\Omega))$$

with $\partial_t \boldsymbol{u} \in L^{p'}(0,T;\mathbb{X}^p_T(\Omega)')$ *satisfies the following equality: for a.e.* $t \in (0,T)$,

$$\langle \partial_t \boldsymbol{u}(t), \boldsymbol{w}\rangle_{\mathbb{X}^p_T(\Omega)',\mathbb{X}^p_T(\Omega)} + \int_\Omega S_s(x,t,|\operatorname{curl}\boldsymbol{u}(t)|^2)\operatorname{curl}\boldsymbol{u}(t)\cdot\operatorname{curl}\boldsymbol{w}dx$$
$$= \int_\Omega \boldsymbol{F}(t)\cdot\boldsymbol{w}dx + \langle \boldsymbol{h}(t)\times\boldsymbol{n},\boldsymbol{w}\rangle_\Gamma \quad (3.1)$$

for all $\boldsymbol{w} \in \mathbb{X}^p_T(\Omega)$, *where*

$$\langle \boldsymbol{h}(t)\times\boldsymbol{n},\boldsymbol{w}\rangle_\Gamma = \langle \boldsymbol{h}(t)\times\boldsymbol{n},\boldsymbol{w}\rangle_{\boldsymbol{W}^{-1/p',p'}(\Gamma),\boldsymbol{W}^{1-1/p,p}(\Gamma)},$$

and $\boldsymbol{u}(0) = \boldsymbol{u}_0$.

To solve (1.6), the next lemma takes an important role. In order to do so, we define a nonlinear operator

$$A(t): \mathbb{V}^p_T(\Omega) \to \mathbb{X}^p_T(\Omega)'$$

by

$$\langle A(t)\boldsymbol{u},\boldsymbol{v}\rangle = \int_\Omega S_s(x,t,|\operatorname{curl}\boldsymbol{u}|^2)\operatorname{curl}\boldsymbol{u}\cdot\operatorname{curl}\boldsymbol{v}dx \quad (3.2)$$

for $\boldsymbol{u} \in \mathbb{V}^p_T(\Omega), \boldsymbol{v} \in \mathbb{X}^p_T(\Omega)$ and define a functional $\boldsymbol{L}(t) \in \mathbb{X}^p_T(\Omega)'$ by

$$\langle \boldsymbol{L}(t),\boldsymbol{v}\rangle = \int_\Omega \boldsymbol{F}(t)\cdot\boldsymbol{v}dx + \langle \boldsymbol{h}(t)\times\boldsymbol{n},\boldsymbol{v}\rangle_\Gamma \text{ for } \boldsymbol{v} \in \mathbb{X}^p_T(\Omega). \quad (3.3)$$

Then we have

Lemma 3.2. *Under the above notations, the system*

$$\begin{cases} \partial_t \boldsymbol{u} + A(t)\boldsymbol{u} = \boldsymbol{L}(t), \\ \boldsymbol{u}(0) = \boldsymbol{u}_0 \end{cases} \tag{3.4}$$

has a unique solution $\boldsymbol{u} \in L^p(0,T;\mathbb{V}_T^p(\Omega)) \cap C([0,T];\boldsymbol{L}^2_\sigma(\Omega))$ *with* $\partial_t \boldsymbol{u} \in L^{p'}(0,T;\mathbb{V}_T^p(\Omega)')$. *The first equation of* (3.4) *is satisfied in the sense of* $L^{p'}(0,T;\mathbb{V}_T^p(\Omega)')$.

Proof. From (2.3a) and the Hölder inequality, we have

$$\begin{aligned} |\langle A(t)\boldsymbol{u}, \boldsymbol{v}\rangle| &\leq \Lambda \|\operatorname{curl} \boldsymbol{u}\|_{\boldsymbol{L}^p(\Omega)}^{p-1} \|\operatorname{curl} \boldsymbol{v}\|_{\boldsymbol{L}^p(\Omega)} \\ &\leq \Lambda \|\boldsymbol{u}\|_{\mathbb{V}_T^p(\Omega)}^{p-1} \|\boldsymbol{v}\|_{\mathbb{X}_T^p(\Omega)} \text{ for all } \boldsymbol{v} \in \mathbb{X}_T^p(\Omega). \end{aligned}$$

Hence

$$A(t)\boldsymbol{u} \in \mathbb{X}_T^p(\Omega)'. \tag{3.5}$$

In particular, if $\boldsymbol{v} \in \mathbb{V}_T^p(\Omega)$, then we have

$$|\langle A(t)\boldsymbol{u}, \boldsymbol{v}\rangle| \leq \Lambda \|\boldsymbol{u}\|_{\mathbb{V}_T^p(\Omega)}^{p-1} \|\boldsymbol{v}\|_{\mathbb{V}_T^p(\Omega)}.$$

Hence $A(t)\boldsymbol{u} \in \mathbb{V}_T^p(\Omega)'$ and

$$\|A(t)\boldsymbol{u}\|_{\widetilde{\mathbb{V}}_T^p(\Omega)'} \leq \Lambda \|\boldsymbol{u}\|_{\mathbb{V}_T^p(\Omega)}^{p-1}.$$

On the other hand, we have

$$|\langle \boldsymbol{L}(t), \boldsymbol{v}\rangle| \leq C(\|\boldsymbol{F}(t)\|_{\boldsymbol{L}^{p'}(\Omega)} + \|\boldsymbol{h}(t) \times \boldsymbol{n}\|_{\boldsymbol{W}^{-1/p',p'}(\Gamma)}) \|\boldsymbol{v}\|_{\mathbb{X}_T^p(\Omega)}.$$

Therefore,

$$\boldsymbol{L}(t) \in \mathbb{X}_T^p(\Omega)' \hookrightarrow \mathbb{V}_T^p(\Omega)'. \tag{3.6}$$

Clearly $A(t)$ is hemi-continuous, i.e., for any $\boldsymbol{v}, \boldsymbol{w}, \boldsymbol{\varphi} \in \mathbb{V}_T^p(\Omega)$ and $\tau \in \mathbb{R}$, $\langle A(t)(\boldsymbol{v} + \tau \boldsymbol{w}), \boldsymbol{\varphi}\rangle$ is continuous in τ. From (2.3a), we have

$$\langle A(t)\boldsymbol{u}, \boldsymbol{u}\rangle \geq \lambda \|\operatorname{curl} \boldsymbol{u}\|_{\boldsymbol{L}^p(\Omega)} = \lambda \|\boldsymbol{u}\|_{\mathbb{V}_T^p(\Omega)} \text{ for } \boldsymbol{u} \in \mathbb{V}_T^p(\Omega).$$

Thus $A(t)$ is coercive on $\mathbb{V}_T^p(\Omega)$, therefore we can apply the celebrated result of Lions [9, Chapter 2, Theorem 1.2] (cf. Zheng [15, Theorem 3.2.1]). That is, (3.4) has a unique solution $\boldsymbol{u} \in L^p(0,T;\mathbb{V}_T^p(\Omega)) \cap C([0,T];\boldsymbol{L}^2_\sigma(\Omega))$ with $\partial_t \boldsymbol{u} \in L^{p'}(0,T;\mathbb{V}_T^p(\Omega)')$. □

Applying Lemma 3.2, we obtain the following proposition.

Proposition 3.3. *Let* $\boldsymbol{F} \in L^{p'}(0,T;\boldsymbol{L}^{p'}(\Omega))$ *satisfy* $\operatorname{div}\boldsymbol{F} = 0$ *in* Ω_T, $\boldsymbol{h}\times\boldsymbol{n} \in L^{p'}(0,T;\boldsymbol{W}^{-1/p',p'}(\Gamma))$ *and* $\boldsymbol{u}_0 \in \boldsymbol{L}^2_\sigma(\Omega)$. *Assume that for a.e.* $t \in (0,T)$,

$$\boldsymbol{F}(t)\cdot\boldsymbol{n} = \operatorname{div}_\Gamma(\boldsymbol{h}(t)\times\boldsymbol{n}) \text{ on } \Gamma, \tag{3.7}$$

and

$$\int_\Omega \boldsymbol{F}(t)\cdot\boldsymbol{v}dx + \langle \boldsymbol{h}(t)\times\boldsymbol{n},\boldsymbol{v}\rangle_\Gamma = 0 \text{ for all } \boldsymbol{v}\in\mathbb{K}^p_T(\Omega). \tag{3.8}$$

Then (1.6) *has a unique weak solution* $\boldsymbol{u}\in L^p(0,T;\mathbb{V}^p_T(\Omega))\cap C([0,T];\boldsymbol{L}^2_\sigma(\Omega))$ *with* $\partial_t\boldsymbol{u}\in L^{p'}(0,T;\mathbb{V}^p_T(\Omega)')$. *Furthermore, there exists a constant* $C>0$ *depending only on* p,Ω *and* λ *such that*

$$\begin{aligned}&\|\boldsymbol{u}\|^2_{L^\infty(0,T;\boldsymbol{L}^2(\Omega))} + \|\operatorname{curl}\boldsymbol{u}\|^p_{L^p(0,T;\boldsymbol{L}^p(\Omega))}\\ &\qquad \le C(\|\boldsymbol{F}\|^{p'}_{L^{p'}(0,T;\boldsymbol{L}^{p'}(\Omega))} + \|\boldsymbol{h}\times\boldsymbol{n}\|^{p'}_{L^{p'}(0,T:\boldsymbol{W}^{-1/p',p'}(\Gamma))} + \|\boldsymbol{u}_0\|^2_{\boldsymbol{L}^2(\Omega)}).\end{aligned}$$

Proof. Define an operator $A(t)$ and a functional $\boldsymbol{L}(t)$ as in (3.2) and (3.3), respectively Then, from Lemma 3.2, (3.4) has a unique weak solution $\boldsymbol{u}\in L^p(0,T;\mathbb{V}^p_T(\Omega))\cap C([0,T];\boldsymbol{L}^2_\sigma(\Omega))$ with $\partial_t\boldsymbol{u}\in L^{p'}(0,T;\mathbb{V}^p_T(\Omega)')$. However, since $A(t)\boldsymbol{u}\in\mathbb{X}^p_T(\Omega)'$ and $\boldsymbol{L}(t)\in\mathbb{X}^p_T(\Omega)'$ from (3.5) and (3.6), we see that $\partial_t\boldsymbol{u}\in L^{p'}(0,T;\mathbb{X}^p_T(\Omega)')$. Thus the first equation of (3.4) holds in $L^{p'}(0,T;\mathbb{X}^p_T(\Omega)')$. Therefore, for a.e. $t\in(0,T)$,

$$\begin{aligned}&\langle\partial_t\boldsymbol{u}(t),\boldsymbol{v}\rangle_{\mathbb{X}^p_T(\Omega)',\mathbb{X}^p_T(\Omega)} + \int_\Omega S_s(x,t,|\operatorname{curl}\boldsymbol{u}(t)|^2)\operatorname{curl}\boldsymbol{u}(t)\cdot\operatorname{curl}\boldsymbol{v}dx\\ &\qquad = \int_\Omega \boldsymbol{F}(t)\cdot\boldsymbol{v}dx + \langle\boldsymbol{h}(t)\times\boldsymbol{n},\boldsymbol{v}\rangle_\Gamma \text{ for all } \boldsymbol{v}\in\mathbb{V}^p_T(\Omega).\end{aligned} \tag{3.9}$$

Since $\boldsymbol{D}(\Omega)$ is not contained in $\mathbb{V}^p_T(\Omega)$, where $\boldsymbol{D}(\Omega)$ is the space of $\boldsymbol{C}^\infty$ functions with compact support in Ω, we can not show that (3.9) implies (1.6) in the distribution sense. To overcome it, we show that we can replace the space $\mathbb{V}^p_T(\Omega)$ of test functions of (3.9) with $\mathbb{X}^p_T(\Omega)$. For any $\boldsymbol{w}\in\mathbb{X}^p_T(\Omega)$, we consider the following Neumann problem.

$$\begin{cases}\Delta\theta = \operatorname{div}\boldsymbol{w} & \text{in } \Omega,\\ \frac{\partial\theta}{\partial\boldsymbol{n}} = 0 & \text{on } \Gamma.\end{cases} \tag{3.10}$$

Here and hereafter, we denote the Laplacian and the gradient operator with respect to the space variable x by Δ and ∇, respectively. Since $\boldsymbol{w} \in \mathbb{X}_T^p(\Omega)$ and $\operatorname{div} \boldsymbol{w} \in L^p(\Omega)$, and from the divergence theorem, we have

$$\int_\Omega \operatorname{div} \boldsymbol{w} dx = \langle \boldsymbol{w} \cdot \boldsymbol{n}, 1 \rangle_\Gamma = 0.$$

Therefore, (3.10) has a unique solution $\theta \in W^{2,p}(\Omega)$, up to an additive constant. Since $\nabla\theta \in \boldsymbol{W}^{1,p}(\Omega)$ and $\boldsymbol{n} \cdot \nabla\theta = 0$ on Γ, we see that $\nabla\theta \in \mathbb{X}_T^p(\Omega)$. Define

$$\boldsymbol{v} = \boldsymbol{w} - \nabla\theta - \sum_{j=1}^{J} \langle (\boldsymbol{w} - \nabla\theta) \cdot \boldsymbol{n}, 1 \rangle_{\Sigma_j} \widetilde{\nabla} q_j^T.$$

Since $\operatorname{div} \boldsymbol{v} = \operatorname{div} \boldsymbol{w} - \Delta\theta = 0$ in Ω, $\boldsymbol{v} \cdot \boldsymbol{n} = 0$ on Γ and $\langle \boldsymbol{v} \cdot \boldsymbol{n}, 1 \rangle_{\Sigma_j} = 0$ for $j = 1, \ldots, J$, we see that $\boldsymbol{v} \in \mathbb{V}_T^p(\Omega)$ and $\operatorname{curl} \boldsymbol{v} = \operatorname{curl} \boldsymbol{w}$ in Ω. Here we use the following lemma whose proof is given in the Appendix.

Lemma 3.4. *Assume that $\boldsymbol{z} \in \boldsymbol{L}^{p'}(\Omega)$ with $\operatorname{curl} \boldsymbol{z} \in \mathbb{X}_T^p(\Omega)'$. Then $\boldsymbol{z} \times \boldsymbol{n} \in \boldsymbol{W}^{-1/p',p'}(\Gamma)$ is well defined, and the following Green's formula holds.*

$$\langle \boldsymbol{z} \times \boldsymbol{n}, \boldsymbol{\varphi} \rangle_\Gamma = \langle \operatorname{curl} \boldsymbol{z}, \boldsymbol{\varphi} \rangle_{\mathbb{X}_T^p(\Omega)', \mathbb{X}_T^p(\Omega)} - \int_\Omega \boldsymbol{z} \cdot \operatorname{curl} \boldsymbol{\varphi} dx \qquad (3.11)$$

for all $\boldsymbol{\varphi} \in \mathbb{X}_T^p(\Omega)$.

From Lemma 3.4 and (3.7), we have

$$\begin{aligned}
&\langle \partial_t \boldsymbol{u}(t), \nabla\theta \rangle_{\mathbb{X}_T^p(\Omega)', \mathbb{X}_T^p(\Omega)} \\
&= \int_\Omega \boldsymbol{F}(t) \cdot \nabla\theta dx - \langle \operatorname{curl} [S_s(x,t,|\operatorname{curl} \boldsymbol{u}(t)|^2) \operatorname{curl} \boldsymbol{u}(t)], \nabla\theta \rangle_{\mathbb{X}_T^p(\Omega)', \mathbb{X}_T^p(\Omega)} \\
&= \langle \boldsymbol{F}(t) \cdot \boldsymbol{n}, \theta \rangle_\Gamma - \langle S_s(x,t,|\operatorname{curl} \boldsymbol{u}(t)|^2) \operatorname{curl} \boldsymbol{u}(t) \times \boldsymbol{n}, \nabla\theta \rangle_\Gamma \\
&= \langle \boldsymbol{F}(t) \cdot \boldsymbol{n} - \operatorname{div}_\Gamma (\boldsymbol{h}(t) \times \boldsymbol{n}), \theta \rangle_\Gamma = 0.
\end{aligned}$$

Similarly, from (3.8),

$$\langle \partial_t \boldsymbol{u}(t), \widetilde{\nabla} q_j^T \rangle_{\mathbb{X}_T^p(\Omega)', \mathbb{X}_T^p(\Omega)} = \int_\Omega \boldsymbol{F}(t) \cdot \widetilde{\nabla} q_j^T dx + \langle \boldsymbol{h}(t) \times \boldsymbol{n}, \widetilde{\nabla} q_j^T \rangle_\Gamma = 0.$$

Thus we have

$$\langle \partial_t \boldsymbol{u}(t), \boldsymbol{v} \rangle_{\mathbb{X}_T^p(\Omega)', \mathbb{X}_T^p(\Omega)} = \langle \partial_t \boldsymbol{u}(t), \boldsymbol{w} \rangle_{\mathbb{X}_T^p(\Omega)', \mathbb{X}_T^p(\Omega)}.$$

Also we have

$$\int_\Omega \boldsymbol{F}(t)\cdot\nabla\theta dx + \langle \boldsymbol{h}(t)\times\boldsymbol{n},\nabla\theta\rangle_\Gamma = \langle \boldsymbol{F}(t)\cdot\boldsymbol{n} - \mathrm{div}_\Gamma(\boldsymbol{h}(t)\times\boldsymbol{n}),\theta\rangle_\Gamma = 0$$

and from (3.8),

$$\int_\Omega \boldsymbol{F}(t)\cdot\widetilde{\nabla} q_j^T dx + \langle \boldsymbol{h}(t)\times\boldsymbol{n},\widetilde{\nabla} q_j^T\rangle_\Gamma = 0.$$

Therefore, we get (3.1) for all $\boldsymbol{w}\in\mathbb{X}_T^p(\Omega)$.

We show the uniqueness of a weak solution. Let $\boldsymbol{u}_i$ $(i=1,2)$ be two weak solutions of (1.6). Taking $\boldsymbol{u}_1-\boldsymbol{u}_2$ as a test function of (3.1) and integrating over $(0,t)$, we have

$$\begin{aligned}\frac{1}{2}\int_\Omega |\boldsymbol{u}_1(t)-\boldsymbol{u}_2(t)|^2dx + \int_0^t\int_\Omega (S_s(x,\tau,|\mathrm{curl}\,\boldsymbol{u}_1(\tau)|^2)\mathrm{curl}\,\boldsymbol{u}_1(\tau)\\ - S_s(x,\tau,|\mathrm{curl}\,\boldsymbol{u}_2(\tau)|^2)\mathrm{curl}\,\boldsymbol{u}_2(\tau))\cdot\mathrm{curl}\,(\boldsymbol{u}_1(\tau)-\boldsymbol{u}_2(\tau))dxd\tau = 0.\end{aligned}$$

By the strict monotonicity (Lemma 2.3) of S_s, we have $\boldsymbol{u}_1=\boldsymbol{u}_2$ in $\mathbb{V}_T^p(\Omega)$.

Finally we show the estimate. Taking $\boldsymbol{u}$ as a test function of (3.1) and integrating over $(0,t)$, by the Hölder and the Young inequalities, we have

$$\begin{aligned}&\frac{1}{2}\|\boldsymbol{u}(t)\|^2_{\boldsymbol{L}^2(\Omega)} + \lambda\|\mathrm{curl}\,\boldsymbol{u}\|^p_{\boldsymbol{L}^p(\Omega_t)}\\ &\quad\le \int_0^t\int_\Omega \boldsymbol{F}(\tau)\cdot\boldsymbol{u}(\tau)dxd\tau + \int_0^t \langle\boldsymbol{h}(\tau)\times\boldsymbol{n},\boldsymbol{u}(\tau)\rangle_\Gamma d\tau + \frac{1}{2}\|\boldsymbol{u}_0\|^2_{\boldsymbol{L}^2(\Omega)}\\ &\quad\le C\int_0^t (\|\boldsymbol{F}(\tau)\|^{p'}_{\boldsymbol{L}^{p'}(\Omega)} + \|\boldsymbol{h}(\tau)\times\boldsymbol{n}\|^{p'}_{\boldsymbol{W}^{-1/p',p'}(\Gamma)})d\tau\\ &\qquad + \frac{\lambda}{2}\|\mathrm{curl}\,\boldsymbol{u}\|^p_{\boldsymbol{L}^p(\Omega_t)} + \frac{1}{2}\|\boldsymbol{u}_0\|^2_{\boldsymbol{L}^2(\Omega)}.\end{aligned}$$

Therefore, there exists a constant $C>0$ depending only on p,Ω and λ such that

$$\begin{aligned}&\|\boldsymbol{u}(t)\|^2_{\boldsymbol{L}^2(\Omega)} + \|\mathrm{curl}\,\boldsymbol{u}\|^p_{\boldsymbol{L}^p(\Omega_t)}\\ &\quad\le C(\|\boldsymbol{F}\|^{p'}_{\boldsymbol{L}^{p'}(0,T;\boldsymbol{L}^{p'}(\Omega))} + \|\boldsymbol{h}\times\boldsymbol{n}\|^{p'}_{L^{p'}(0,T;\boldsymbol{W}^{-1/p',p'}(\Gamma))} + \|\boldsymbol{u}_0\|^2_{\boldsymbol{L}^2(\Omega)}).\end{aligned}$$

Taking the sup over $(0,T)$, we get the estimate. This completes the proof of Proposition 3.3. □

Remark 3.5. *In [10], the authors considered the similar problem in the case where Ω is simply connected. In this case, the fifth conditions of* (1.6) *are unnecessary. Furthermore, since they took $\mathbb{V}_T^p(\Omega)$ as the space of test functions in* (3.1)*, a weak solution is not a solution of* (1.6) *in the distribution sense. However, as we showed that we can take $\mathbb{X}_T^p(\Omega)$ as the space of test functions, our weak solution is a solution of* (1.6) *in the distribution sense. Hence we must assume that the compatibility conditions* $\operatorname{div} \boldsymbol{F} = 0$ *in* Ω_T *and* (3.7)*.* (3.8) *is the compatibility condition for the stationary problem (cf. Aramaki [5]).*

4. THE MAXWELL-STOKES TYPE PROBLEM

In this section, we consider the Maxwell-Stokes system (1.7). We give the notion of a weak solution for the system (1.7).

Definition 4.1. $(\boldsymbol{u}, \pi)$ *is a weak solution of* (1.7)*, if*

$$(\boldsymbol{u}, \pi) \in \left(L^p(0,T;\mathbb{V}_T^p(\Omega)) \cap C([0,T];\boldsymbol{L}_\sigma^2(\Omega))\right) \times L^{p'}(0,T;W^{1,p'}(\Omega)/\mathbb{R})$$

with $\partial_t \boldsymbol{u} \in L^{p'}(0,T;\mathbb{X}_T^p(\Omega)')$ *satisfies that, for a.e.* $t \in (0,T)$*,*

$$\begin{aligned} &\langle \partial_t \boldsymbol{u}(t), \boldsymbol{v} \rangle_{\mathbb{X}_T^p(\Omega)', \mathbb{X}_T^p(\Omega)} + \int_\Omega S_s(x,t,|\operatorname{curl} \boldsymbol{u}(t)|^2) \operatorname{curl} \boldsymbol{u}(t) \cdot \operatorname{curl} \boldsymbol{v} dx \\ &+ \int_\Omega \nabla \pi(t) \cdot \boldsymbol{v} dx = \int_\Omega \boldsymbol{f}(t) \cdot \boldsymbol{v} dx + \langle \boldsymbol{h}(t) \times \boldsymbol{n}, \boldsymbol{v} \rangle_\Gamma \text{ for all } \boldsymbol{v} \in \mathbb{X}_T^p(\Omega) \end{aligned} \tag{4.1}$$

and $\boldsymbol{u}(0) = \boldsymbol{u}_0$.

We obtain the following theorem.

Theorem 4.2. *Assume that* $6/5 \le p < \infty$, $S(x,t,s)$ *satisfies the structure conditions* (2.3a)-(2.3c)*, and that* $\boldsymbol{f} \in L^{p'}(0,T;\boldsymbol{L}^{p'}(\Omega))$ *with* $\operatorname{div} \boldsymbol{f} \in L^{p'}(0,T;\boldsymbol{L}^{p'}(\Omega))$, $\boldsymbol{h} \times \boldsymbol{n} \in \boldsymbol{L}^{p'}(0,T;\boldsymbol{W}^{-1/p',p'}(\Gamma))$ *with* $\operatorname{div}_\Gamma(\boldsymbol{h} \times \boldsymbol{n}) \in \boldsymbol{L}^{p'}(0,T;W^{-1/p',p'}(\Gamma))$ *and* $\boldsymbol{u}_0 \in \boldsymbol{L}_\sigma^2(\Omega)$*. Furthermore, we assume that for a.e.* $t \in (0,T)$*,*

$$\int_\Omega \boldsymbol{f}(t) \cdot \boldsymbol{v} dx + \langle \boldsymbol{h}(t) \times \boldsymbol{n}, \boldsymbol{v} \rangle_\Gamma = 0 \text{ for all } \boldsymbol{v} \in \mathbb{K}_T^p(\Omega). \tag{4.2}$$

Then (1.7) *has a unique weak solution* $(\boldsymbol{u},\pi)$, *and there exists a constant* $C>0$ *depending only on* p,λ,Λ *and* Ω *such that*

$$\begin{aligned}&\|\boldsymbol{u}\|^2_{L^\infty(0,T;\boldsymbol{L}^2(\Omega))}+\|\mathrm{curl}\,\boldsymbol{u}\|^p_{\boldsymbol{L}^p(\Omega_T)}+\|\pi\|^{p'}_{L^{p'}(0,T;W^{1,p'}(\Omega)/\mathbb{R})}\\&\le C(\|\boldsymbol{f}\|^{p'}_{\boldsymbol{L}^{p'}(0,T;\boldsymbol{L}^{p'}(\Omega))}+\|\mathrm{div}\,\boldsymbol{f}\|^{p'}_{L^{p'}(0,T;L^{p'}(\Omega))}+\|\boldsymbol{h}\times\boldsymbol{n}\|^{p'}_{\boldsymbol{L}^{p'}(0,T;\boldsymbol{W}^{-1/p',p'}(\Gamma))}\\&\qquad+\|\mathrm{div}_\Gamma(\boldsymbol{h}\times\boldsymbol{n})\|^{p'}_{\boldsymbol{L}^{p'}(0,T;\boldsymbol{W}^{-1/p',p'}(\Gamma))}+\|\boldsymbol{u}_0\|^2_{\boldsymbol{L}^2(\Omega)}).\end{aligned}\tag{4.3}$$

Proof. First, for a.e. $t\in(0,T)$, we consider the following Neumann problem.

$$\begin{cases}\Delta\pi=\mathrm{div}\,\boldsymbol{f}(t) & \text{in }\Omega,\\ \frac{\partial\pi}{\partial\boldsymbol{n}}=\boldsymbol{f}(t)\cdot\boldsymbol{n}-\mathrm{div}_\Gamma(\boldsymbol{h}(t)\times\boldsymbol{n}) & \text{on }\Gamma.\end{cases}\tag{4.4}$$

Since

$$\begin{aligned}\langle\mathrm{div}_\Gamma(\boldsymbol{h}(t)\times\boldsymbol{n}),1\rangle_\Gamma&=\langle\mathrm{div}_\Gamma(\boldsymbol{h}(t)\times\boldsymbol{n}),1\rangle_{W^{-1-1/p',p'}(\Gamma),W^{2-1/p,p}(\Gamma)}\\&=\langle\boldsymbol{h}(t)\times\boldsymbol{n},\nabla1\rangle_\Gamma=0,\end{aligned}$$

the data of (4.4) satisfy the compatibility condition. Hence, according to Aramaki [3], there exists a unique weak solution $\pi(t)\in W^{1,p'}(\Omega)/\mathbb{R}$, and there exists a constant $C>0$ depending only on p' and Ω such that

$$\begin{aligned}&\|\pi(t)\|_{W^{1,p'}(\Omega)/\mathbb{R}}\\&\le C(\|\boldsymbol{f}(t)\|_{\boldsymbol{L}^{p'}(\Omega)}+\|\mathrm{div}\,\boldsymbol{f}(t)\|_{L^{p'}(\Omega)}+\|\mathrm{div}_\Gamma(\boldsymbol{h}(t)\times\boldsymbol{n})\|_{\boldsymbol{W}^{-1/p',p'}(\Gamma)}).\end{aligned}\tag{4.5}$$

Here we apply the following Lusin theorem: a function $f(t)$ on a Lebesgue measurable set $E\subset\mathbb{R}^d$ is Lebesgue measurable, if and only if for any $\varepsilon>0$, there exists a closed subset $F\subset E$ such that $\mu(E\setminus F)<\varepsilon$ and $f(t)$ is continuous on F, where μ denotes the Lebesgue measure on $\mathbb{R}^d$.

Since

$$\begin{aligned}\|\pi(t)-\pi(t')\|_{W^{1,p'}(\Omega)/\mathbb{R}}&\le C(\|\boldsymbol{f}(t)-\boldsymbol{f}(t')\|_{\boldsymbol{L}^{p'}(\Omega)}+\|\mathrm{div}\,(\boldsymbol{f}(t)-\boldsymbol{f}(t'))\|_{\boldsymbol{L}^{p'}(\Omega)}\\&\qquad+\|\mathrm{div}_\Gamma((\boldsymbol{h}(t)-\boldsymbol{h}(t'))\times\boldsymbol{n})\|_{\boldsymbol{W}^{-1/p',p'}(\Gamma)}),\end{aligned}$$

we can see that $\|\pi(t)\|_{W^{1,p'}(\Omega)/\mathbb{R}}$ is Lebesgue measurable on $E=(0,T)\subset\mathbb{R}$, and so we can derive $\pi\in L^{p'}(0,T;W^{1,p'}(\Omega)/\mathbb{R})$, and

$$\begin{aligned}\|\pi\|_{L^{p'}(0,T;W^{1,p'}(\Omega)/\mathbb{R})}&\le C(\|\boldsymbol{f}\|_{L^{p'}(0,T;\boldsymbol{L}^{p'}(\Omega))}\\&\qquad+\|\mathrm{div}\,\boldsymbol{f}\|_{L^{p'}(0,T;L^{p'}(\Omega))}+\|\mathrm{div}_\Gamma(\boldsymbol{h}\times\boldsymbol{n})\|_{L^{p'}(0,T;W^{-1/p',p'}(\Gamma))}).\end{aligned}\tag{4.6}$$

If we put $\boldsymbol{F} = \boldsymbol{f} - \nabla\pi \in L^{p'}(0,T;\boldsymbol{L}^{p'}(\Omega))$, then we have $\operatorname{div}\boldsymbol{F}(t) = \operatorname{div}\boldsymbol{f}(t) - \Delta\pi(t) = 0$ in Ω and $\boldsymbol{F}\cdot\boldsymbol{n} = \boldsymbol{f}\cdot\boldsymbol{n} - \frac{\partial\pi}{\partial\boldsymbol{n}} = \operatorname{div}_\Gamma(\boldsymbol{h}\times\boldsymbol{n})$ on Γ. For any $\boldsymbol{v}\in\mathbb{K}_T^p(\Omega)$, since

$$\int_\Omega \nabla\pi\cdot\boldsymbol{v}dx = \int_\Gamma \pi(\boldsymbol{v}\cdot\boldsymbol{n})dS - \int_\Omega \pi\operatorname{div}\boldsymbol{v}dx = 0,$$

from (4.2), where dS denotes the surface measure on Γ, we have, for a.e. $t\in(0,T)$,

$$\int_\Omega \boldsymbol{F}(t)\cdot\boldsymbol{v}dx + \langle\boldsymbol{h}(t)\times\boldsymbol{n},\boldsymbol{v}\rangle_\Gamma = \int_\Omega \boldsymbol{f}(t)\cdot\boldsymbol{v}dx + \langle\boldsymbol{h}(t)\times\boldsymbol{n},\boldsymbol{v}\rangle_\Gamma = 0$$

for all $\boldsymbol{v}\in\mathbb{K}_T^p(\Omega)$. Therefore, if we use the result of Proposition 3.3, then (1.7) has a weak solution

$$(\boldsymbol{u},\pi)\in\left(L^p(0,T;\mathbb{V}_T^p(\Omega))\cap C([0,T];\boldsymbol{L}_\sigma^2(\Omega))\right)\times L^{p'}(0,T;W^{1,p'}(\Omega)/\mathbb{R})$$

with $\partial_t\boldsymbol{u}\in\boldsymbol{L}^{p'}(0,T;\mathbb{X}_T^p(\Omega)')$. Since form (4.5),

$$\begin{aligned}\|\boldsymbol{F}\|_{L^{p'}(0,T;\boldsymbol{L}^{p'}(\Omega))} &\le \|\boldsymbol{f}\|_{L^{p'}(0,T;\boldsymbol{L}^{p'}(\Omega))} + \|\nabla\pi\|_{L^{p'}(0,T;\boldsymbol{L}^{p'}(\Omega))}\\ &\le C(\|\boldsymbol{f}\|_{L^{p'}(0,T;\boldsymbol{L}^{p'}(\Omega))} + \|\operatorname{div}\boldsymbol{f}\|_{L^{p'}(0,T;\boldsymbol{L}^{p'}(\Omega))}\\ &\qquad + \|\operatorname{div}_\Gamma(\boldsymbol{h}\times\boldsymbol{n})\|_{L^{p'}(0,T;W^{-1/p',p'}(\Gamma))}).\end{aligned}$$

Hence from the estimate of Proposition 3.3, the estimate (4.3) holds.

Finally we show the uniqueness of the weak solution. Let $(\boldsymbol{u}_1,\pi_1),(\boldsymbol{u}_2,\pi_2)$ be two weak solutions of (1.7). Taking $\boldsymbol{u}_1-\boldsymbol{u}_2$ as a test function of (4.1), we have

$$\begin{aligned}&\langle\partial_t(\boldsymbol{u}_1(t)-\boldsymbol{u}_2(t)),\boldsymbol{u}_1(t)-\boldsymbol{u}_2(t)\rangle_{\mathbb{X}_T^p(\Omega)',\mathbb{X}_T^p(\Omega)}\\ &+\int_\Omega (S_s(x,t,|\operatorname{curl}\boldsymbol{u}_1(t)|^2)\operatorname{curl}\boldsymbol{u}_1(t) - S_s(x,t,|\operatorname{curl}\boldsymbol{u}_2(t)|^2)\operatorname{curl}\boldsymbol{u}_2(t))\cdot\operatorname{curl}(\boldsymbol{u}_1(t)-\boldsymbol{u}_2(t))dx\\ &\qquad + \int_\Omega \nabla(\pi_1(t)-\pi_2(t))\cdot(\boldsymbol{u}_1(t)-\boldsymbol{u}_2(t))dx = 0.\end{aligned}$$

Since $\boldsymbol{u}_1(t)-\boldsymbol{u}_2(t)\in\mathbb{V}_T^p(\Omega)$, the last integral is equal to zero. Integrating this equality over $(0,t)$ and using $\boldsymbol{u}_1(0)-\boldsymbol{u}_2(0)=\boldsymbol{0}$, we have

$$\begin{aligned}\frac{1}{2}\int_\Omega |\boldsymbol{u}_1(t)-\boldsymbol{u}_2(t)|^2dx + \int_0^t\int_\Omega (S_s(x,\tau,|\operatorname{curl}\boldsymbol{u}_1(\tau)|^2)\operatorname{curl}\boldsymbol{u}_1(\tau)\\ - S_s(x,\tau,|\operatorname{curl}\boldsymbol{u}_2(\tau)|^2)\operatorname{curl}\boldsymbol{u}_2(\tau))\cdot\operatorname{curl}(\boldsymbol{u}_1(\tau)-\boldsymbol{u}_2(\tau))dxd\tau = 0.\end{aligned}$$

By the strict monotonicity of S_t (Lemma 2.3), we have $\boldsymbol{u}_1 = \boldsymbol{u}_2$. Moreover, taking $\boldsymbol{v} \in \boldsymbol{D}(\Omega)$ as a test function of (4.1), we have

$$\int_\Omega \nabla(\pi_1 - \pi_2) \cdot \boldsymbol{v} dx = 0,$$

so $\nabla(\pi_1 - \pi_2) = 0$ in the distribution sense. Therefore, we have $\pi_1 = \pi_2$ in $L^{p'}(0,T;W^{1,p'}(\Omega)/\mathbb{R})$. □

When $S(x,t,s) = S(x,s)$ is independent of t, we can improve the previous Theorem 4.2, provided that the given data $\boldsymbol{f}$, $\boldsymbol{h}$ and $\boldsymbol{u}_0$ are more regular.

Proposition 4.3. *Let $S(x,s)$ satisfy the structure conditions* (2.3a)-(2.3c) *with the same constants λ and Λ. Assume that $\boldsymbol{f} \in L^\infty(0,T;\boldsymbol{L}^{p'}(\Omega)) \cap \boldsymbol{L}^2(\Omega_T)$ with* $\operatorname{div} \boldsymbol{f} \in L^\infty(0,T;\boldsymbol{L}^{p'}(\Omega)) \cap \boldsymbol{L}^2(\Omega_T)$,

$$\boldsymbol{h} \times \boldsymbol{n} \in L^\infty(0,T;\boldsymbol{W}^{-1/p',p'}(\Gamma)) \cap W^{1,p'}(0,T;\boldsymbol{W}^{-1/p',p'}(\Gamma))$$

with

$$\operatorname{div}_\Gamma(\boldsymbol{h} \times \boldsymbol{n}) \in L^\infty(0,T;W^{-1/p',p'}(\Gamma)) \cap L^2(0,T;H^{-1/2}(\Gamma))$$

and $\boldsymbol{u}_0 \in \mathbb{V}^p_T(\Omega)$. Moreover, we assume that (4.2) *holds. Then the weak solution $(\boldsymbol{u},\pi)$ of* (1.7) *satisfies that $\partial_t \boldsymbol{u} \in \boldsymbol{L}^2(\Omega_T)$,* $\operatorname{curl} \boldsymbol{u} \in L^\infty(0,T;\boldsymbol{L}^p(\Omega))$ *and* $\pi \in L^{p'}(0,T;W^{1,p'}(\Omega)) \cap L^2(0,T;H^1(\Omega))$.

Proof. Using the Galerkin approximation (cf. [15, Chapter 3]), we may choose formally $\partial_t \boldsymbol{u}(t)$ as a test function in (4.1). Integrating the first equality of (4.1) with $\boldsymbol{v} = \partial_t \boldsymbol{u}(t)$ over $(0,t)$ leads to

$$\begin{aligned}\int_0^t \int_\Omega |\partial_\tau \boldsymbol{u}(\tau)|^2 dx d\tau + \int_0^t \int_\Omega S_s(x, |\operatorname{curl} \boldsymbol{u}(\tau)|^2) \operatorname{curl} \boldsymbol{u}(\tau) \cdot \operatorname{curl} \partial_\tau \boldsymbol{u}(\tau) dx d\tau \\ + \int_0^t \int_\Omega \nabla\pi \cdot \partial_\tau \boldsymbol{u}(\tau) dx d\tau = \int_0^t \int_\Omega \boldsymbol{f}(\tau) \cdot \partial_\tau \boldsymbol{u}(\tau) dx d\tau \\ + \int_0^t \langle \boldsymbol{h}(\tau) \times \boldsymbol{n}, \partial_\tau \boldsymbol{u}(\tau) \rangle_\Gamma d\tau. \quad (4.7)\end{aligned}$$

Integrating by parts, we have

$$\begin{aligned}\int_0^t \langle \boldsymbol{h}(\tau) \times \boldsymbol{n}, \partial_\tau \boldsymbol{u}(\tau) \rangle_\Gamma d\tau = \langle \boldsymbol{h}(t) \times \boldsymbol{n}, \boldsymbol{u}(t) \rangle_\Gamma \\ - \langle \boldsymbol{h}(0) \times \boldsymbol{n}, \boldsymbol{u}_0 \rangle_\Gamma - \int_0^t \langle \partial_\tau \boldsymbol{h}(\tau) \times \boldsymbol{n}, \boldsymbol{u}(\tau) \rangle_\Gamma d\tau.\end{aligned}$$

Hence, taking the estimate (4.3) into consideration, we can see that

$$\left|\int_0^t \langle \boldsymbol{h}(\tau)\times \boldsymbol{n}, \partial_\tau \boldsymbol{u}(\tau)\rangle_\Gamma d\tau\right| \le C(\varepsilon)\|\boldsymbol{h}\times\boldsymbol{n}\|^{p'}_{L^\infty(0,T;\boldsymbol{W}^{-1/p',p'}(\Gamma))} + \varepsilon\|\boldsymbol{u}(t)\|^p_{\mathbb{V}^p_T(\Omega)} + C_1,$$

where C_1 is a constant depending only on the data and

$$\left|\int_0^t\int_\Omega \boldsymbol{f}(\tau)\cdot\partial_\tau \boldsymbol{u}(\tau)dxd\tau\right| \le C(\varepsilon)\|\boldsymbol{f}\|^2_{\boldsymbol{L}^2(\Omega_T)} + \varepsilon\int_0^t\int_\Omega |\partial_\tau \boldsymbol{u}(\tau)|^2 dxd\tau$$

for any $\varepsilon > 0$. By the structure condition (2.3a), we have

$$\begin{aligned}
&\int_0^t\int_\Omega S_s(x,|\mathrm{curl}\,\boldsymbol{u}(\tau)|^2)\mathrm{curl}\,\boldsymbol{u}(\tau)\cdot\partial_\tau\boldsymbol{u}(\tau)dxd\tau\\
&\quad=\frac{1}{2}\int_0^t\int_\Omega \frac{d}{d\tau}S(x,|\mathrm{curl}\,\boldsymbol{u}(\tau)|^2)dxd\tau\\
&\quad=\frac{1}{2}\int_\Omega S(x,|\mathrm{curl}\,\boldsymbol{u}(t)|^2)dx-\frac{1}{2}\int_\Omega S(x,\mathrm{curl}\,\boldsymbol{u}_0|^2)dx\\
&\quad\ge\frac{\lambda}{2}\int_\Omega|\mathrm{curl}\,\boldsymbol{u}(t)|^p dx-\frac{\Lambda}{2}\int_\Omega|\mathrm{curl}\,\boldsymbol{u}_0|^p dx.
\end{aligned}$$

Therefore, we have, for any $\varepsilon > 0$,

$$\begin{aligned}
&\int_0^t\int_\Omega|\partial_\tau\boldsymbol{u}(\tau)|^2dxd\tau+\frac{\lambda}{2}\|\mathrm{curl}\,\boldsymbol{u}(t)\|^p_{\boldsymbol{L}^p(\Omega)}\\
&\qquad\le\varepsilon\int_0^t\int_\Omega|\partial_\tau\boldsymbol{u}(\tau)|^2dxd\tau+\left|\int_0^t\int_\Omega\nabla\pi\cdot\partial_\tau\boldsymbol{u}(\tau)dxd\tau\right|+C_3.
\end{aligned}$$

Since, for a.e. $t\in(0,T)$, $\boldsymbol{f}(t)\in\boldsymbol{L}^2(\Omega)$ with $\mathrm{div}\,\boldsymbol{f}(t)\in L^2(\Omega)$ and moreover $\mathrm{div}_\Gamma(\boldsymbol{h}(t)\times\boldsymbol{n})\in H^{-1/2}(\Omega)$, the solution $\pi(t)$ of (4.4) satisfies that $\pi(t)\in H^1(\Omega)$ and

$$\|\pi(t)\|_{H^1(\Omega)}\le C(\Omega)(\|\mathrm{div}\,\boldsymbol{f}(t)\|_{L^2(\Omega)}+\|\boldsymbol{f}(t)\|_{L^2(\Omega)}+\|\mathrm{div}_\Gamma(\boldsymbol{h}(t)\times\boldsymbol{n}\|_{H^{-1/2}(\Gamma)}).$$

Therefore, $\pi\in L^2(0,T;H^1(\Omega))$, and

$$\|\pi\|_{L^2(0,T;H^1(\Omega))}\le C(\Omega)(\|\mathrm{div}\,\boldsymbol{f}\|_{L^2(\Omega_T)}+\|\boldsymbol{f}\|_{L^2(\Omega_T)}+\|\mathrm{div}_\Gamma(\boldsymbol{h}\times\boldsymbol{n})\|_{L^2(0,T;H^{-1/2}(\Gamma)}).$$

Thus, we have

$$\left|\int_0^t\int_\Omega\nabla\pi\cdot\partial_\tau\boldsymbol{u}(\tau)dxd\tau\right|\le C_3(\varepsilon)+\varepsilon\int_0^t\int_\Omega|\partial_\tau\boldsymbol{u}(\tau)|^2dxd\tau.$$

If we choose $\varepsilon > 0$ small enough, we get the conclusion. □

5. Asymptotic Behavior of the Weak Solution as $t \to \infty$

In this section, we show that the weak solution $\boldsymbol{u}(t)$ of (1.7) converges to the solution of a stationary system as $t \to \infty$ in an appropriate space under some conditions. In order to proceed, let a Carathéodory function $S^{(\infty)}(x,s)$ satisfy the same structure conditions (2.3a)-(2.3c) with the same constants. We consider the following stationary system.

$$\begin{cases} \operatorname{curl}[S_s^{(\infty)}(x,|\operatorname{curl}\boldsymbol{u}_\infty|^2)\operatorname{curl}\boldsymbol{u}_\infty] + \nabla\pi_\infty = \boldsymbol{f}_\infty & \text{in } \Omega, \\ \operatorname{div}\boldsymbol{u}_\infty = 0 & \text{in } \Omega, \\ \boldsymbol{u}_\infty \cdot \boldsymbol{n} = 0 & \text{on } \Gamma \\ S^{(\infty)}(x,|\operatorname{curl}\boldsymbol{u}_\infty|^2)\operatorname{curl}\boldsymbol{u}_\infty \times \boldsymbol{n} = \boldsymbol{h}_\infty \times \boldsymbol{n} & \text{on } \Gamma, \\ \langle \boldsymbol{u}_\infty \cdot \boldsymbol{n}, 1\rangle_{\Sigma_j} = 0 & j = 1,\ldots,J. \end{cases} \tag{5.1}$$

We give a proposition which was proved in the previous paper Aramaki [5].

Proposition 5.1. *If $\boldsymbol{f}_\infty \in \boldsymbol{L}^{p'}(\Omega)$ and $\boldsymbol{h}_\infty \times \boldsymbol{n} \in \boldsymbol{W}^{-1/p',p'}(\Gamma)$ satisfies*

$$\int_\Omega \boldsymbol{f}_\infty \cdot \boldsymbol{v} dx + \langle \boldsymbol{h}_\infty \times \boldsymbol{n}, \boldsymbol{v}\rangle_\Gamma = 0 \text{ for all } \boldsymbol{v} \in \mathbb{K}_T^p(\Omega), \tag{5.2}$$

then (5.1) has a unique weak solution $(\boldsymbol{u}_\infty, \pi_\infty) \in \boldsymbol{W}^{1,p}(\Omega) \times L^{p'}(\Omega)/\mathbb{R}$ in the sense of

$$\int_\Omega S_s^{(\infty)}(x,|\operatorname{curl}\boldsymbol{u}_\infty|^2)\operatorname{curl}\boldsymbol{u}_\infty \cdot \operatorname{curl}\boldsymbol{v} dx - \int_\Omega \pi_\infty \operatorname{div}\boldsymbol{v} dx$$
$$= \int_\Omega \boldsymbol{f}_\infty \cdot \boldsymbol{v} dx + \langle \boldsymbol{h}_\infty \times \boldsymbol{n}, \boldsymbol{v}\rangle_\Gamma \text{ for all } \boldsymbol{v} \in \mathbb{X}_T^p(\Omega). \tag{5.3}$$

Assume that $\boldsymbol{f} \in L^\infty(0,\infty;\boldsymbol{L}^{p'}(\Omega))$ with $\operatorname{div}\boldsymbol{f} \in L^\infty(0,\infty;\boldsymbol{L}^{p'}(\Omega))$, $\boldsymbol{h} \times \boldsymbol{n} \in L^\infty(0,\infty;\boldsymbol{W}^{-1/p',p'}(\Gamma))$ with $\operatorname{div}_\Gamma(\boldsymbol{h} \times \boldsymbol{n}) \in L^\infty(0,\infty;W^{-1/p',p'}(\Gamma))$ and $\boldsymbol{u}_0 \in \boldsymbol{L}_\sigma^2(\Omega)$ satisfy that for a.e. $t \in (0,\infty)$,

$$\int_\Omega \boldsymbol{f}(t) \cdot \boldsymbol{v} dx + \langle \boldsymbol{h}(t) \times \boldsymbol{n}, \boldsymbol{v}\rangle_\Gamma = 0 \text{ for all } \boldsymbol{v} \in \mathbb{K}_T^p(\Omega). \tag{5.4}$$

Then from Theorem 4.2, the system (1.7) has a unique weak solution

$$(\boldsymbol{u},\pi) = (\boldsymbol{u}(t),\pi(t)) \in \left(L^p(0,\infty;\mathbb{V}_T^p(\Omega)) \cap C([0,\infty);\boldsymbol{L}_\sigma^2(\Omega))\right) \times L^{p'}(0,\infty;W^{1,p'}(\Omega))$$

with $\partial_t \boldsymbol{u} \in L^{p'}(0,\infty);\mathbb{X}_T^p(\Omega)')$. Since $\pi(t)$ is a solution of (4.4), we have

$$\begin{cases} \Delta(\pi(t)-\pi_\infty) = \operatorname{div}(\boldsymbol{f}(t)-\boldsymbol{f}_\infty) & \text{in } \Omega, \\ \frac{\partial}{\partial \boldsymbol{n}}(\pi(t)-\pi_\infty) = (\boldsymbol{f}(t)-\boldsymbol{f}_\infty)\cdot\boldsymbol{n} - \operatorname{div}_\Gamma((\boldsymbol{h}(t)-\boldsymbol{h}_\infty)\times\boldsymbol{n}) & \text{on } \Gamma. \end{cases}$$

Therefore, it follows from (4.5) that

$$\begin{aligned} \|\pi(t)-\pi_\infty\|_{W^{1,p'}(\Omega)/\mathbb{R}} \le C(p',\Omega)(&\|\operatorname{div}(\boldsymbol{f}(t)-\boldsymbol{f}_\infty)\|_{L^{p'}(\Omega)} \\ &+\|\boldsymbol{f}(t)-\boldsymbol{f}_\infty\|_{L^{p'}(\Omega)} + \|\operatorname{div}_\Gamma((\boldsymbol{h}(t)-\boldsymbol{h}_\infty)\times\boldsymbol{n})\|_{W^{-1/p',p'}(\Gamma)}) \end{aligned} \quad (5.5)$$

For a.e. $t\in(0,\infty)$, taking $\boldsymbol{w}(t) = \boldsymbol{u}(t)-\boldsymbol{u}_\infty$ as a test function of (5.3) and (4.1), we have

$$\begin{aligned} \int_\Omega S_s^{(\infty)}(x,|\operatorname{curl}\boldsymbol{u}_\infty|^2)\operatorname{curl}\boldsymbol{u}_\infty\cdot\operatorname{curl}\boldsymbol{w}(t)dx + \int_\Omega \nabla\pi_\infty\cdot\boldsymbol{w}(t)dx \\ = \int_\Omega \boldsymbol{f}_\infty\cdot\boldsymbol{w}(t)dx + \langle \boldsymbol{h}_\infty\times\boldsymbol{n},\boldsymbol{w}(t)\rangle_\Gamma \end{aligned}$$

and

$$\begin{aligned} \langle \partial_t\boldsymbol{u}(t),\boldsymbol{w}(t)\rangle_{\mathbb{X}_T^p(\Omega)',\mathbb{X}_T^p(\Omega)} + \int_\Omega S_s(x,t,|\operatorname{curl}\boldsymbol{u}(t)|^2)\operatorname{curl}\boldsymbol{u}(t)\cdot\operatorname{curl}\boldsymbol{w}(t)dx \\ + \int_\Omega \nabla\pi(t)\cdot\boldsymbol{w}(t)dx = \int_\Omega \boldsymbol{f}(t)\cdot\boldsymbol{w}(t)dx + \langle \boldsymbol{h}(t)\times\boldsymbol{n},\boldsymbol{w}(t)\rangle_\Gamma. \end{aligned}$$

Therefore, we have

$$\begin{aligned} \langle \partial_t\boldsymbol{w}(t),\boldsymbol{w}(t)\rangle_{\mathbb{X}_T^p(\Omega)',\mathbb{X}_T^p(\Omega)} + \int_\Omega \big(S_s(x,t,|\operatorname{curl}\boldsymbol{u}(t)|^2)\operatorname{curl}\boldsymbol{u}(t) \\ - S_s^{(\infty)}(x,|\operatorname{curl}\boldsymbol{u}_\infty|^2)\operatorname{curl}\boldsymbol{u}_\infty\big)\cdot\operatorname{curl}\boldsymbol{w}(t)dx + \int_\Omega \nabla(\pi(t)-\pi_\infty)\cdot\boldsymbol{w}(t)dx \\ = \int_\Omega(\boldsymbol{f}(t)-\boldsymbol{f}_\infty)\cdot\boldsymbol{w}(t)dx + \langle(\boldsymbol{h}(t)-\boldsymbol{h}_\infty)\times\boldsymbol{n},\boldsymbol{w}(t)\rangle_\Gamma. \end{aligned}$$

We rewrite this equality into the form

$$\begin{aligned} \frac{1}{2}\frac{d}{dt}\int_\Omega |\boldsymbol{w}(t)|^2dx + \int_\Omega \big(S_s(x,t,|\operatorname{curl}\boldsymbol{u}(t)|^2)\operatorname{curl}\boldsymbol{u}(t) \\ - S_s(x,t,|\operatorname{curl}\boldsymbol{u}_\infty|^2)\operatorname{curl}\boldsymbol{u}_\infty\big)\cdot\operatorname{curl}\boldsymbol{w}(t)dx = I_1+I_2+I_3, \end{aligned} \quad (5.6)$$

where

$$\begin{aligned} I_1 &= -\int_\Omega \nabla(\pi(t)-\pi_\infty)\cdot \boldsymbol{w}(t)dx, \\ I_2 &= \int_\Omega (\boldsymbol{f}(t)-\boldsymbol{f}_\infty)\cdot \boldsymbol{w}(t)dx + \langle(\boldsymbol{h}(t)-\boldsymbol{h}_\infty)\times \boldsymbol{n}, \boldsymbol{w}(t)\rangle_\Gamma, \\ I_3 &= \int_\Omega (S_s^{(\infty)}(x,|\mathrm{curl}\,\boldsymbol{u}_\infty|^2)\mathrm{curl}\,\boldsymbol{u}_\infty - S_s(x,t,|\mathrm{curl}\,\boldsymbol{u}_\infty|^2)\mathrm{curl}\,\boldsymbol{u}_\infty)\cdot \mathrm{curl}\,\boldsymbol{w}(t)dx. \end{aligned}$$

We estimate $|I_1|,|I_2|$ and $|I_3|$. From (5.5), for any $\varepsilon>0$,

$$\begin{aligned} |I_1| &\le C(\varepsilon)\|\nabla(\pi(t)-\pi_\infty)\|^{p'}_{\boldsymbol{L}^{p'}(\Omega)} + \varepsilon\|\boldsymbol{w}(t)\|^p_{\boldsymbol{L}^p(\Omega)} \\ &\le C(\varepsilon,p',\Omega)(\|\mathrm{div}(\boldsymbol{f}(t)-\boldsymbol{f}_\infty)\|^{p'}_{L^{p'}(\Omega)} + \|\boldsymbol{f}(t)-\boldsymbol{f}_\infty\|^{p'}_{\boldsymbol{L}^{p'}(\Omega)} \\ &\quad + \|\mathrm{div}_\Gamma((\boldsymbol{h}(t)-\boldsymbol{h}_\infty)\times\boldsymbol{n})\|^{p'}_{W^{-1/p',p'}(\Gamma)}) + \varepsilon\|\boldsymbol{w}(t)\|^p_{\boldsymbol{L}^p(\Omega)}, \end{aligned}$$

$$|I_2| \le C(\varepsilon)(\|\boldsymbol{f}(t)-\boldsymbol{f}_\infty\|^{p'}_{\boldsymbol{L}^{p'}(\Omega)} + \|(\boldsymbol{h}(t)-\boldsymbol{h}_\infty)\times\boldsymbol{n}\|^{p'}_{\boldsymbol{W}^{-1/p',p'}(\Gamma)}) + \varepsilon\|\boldsymbol{w}(t)\|^p_{\boldsymbol{L}^p(\Omega)},$$

and

$$\begin{aligned} |I_3| &\le C(\varepsilon)\|S_s^{(\infty)}(x,\mathrm{curl}\,\boldsymbol{u}_\infty|^2)\mathrm{curl}\,\boldsymbol{u}_\infty - S_s(x,t,|\mathrm{curl}\,\boldsymbol{u}_\infty|^2)\mathrm{curl}\,\boldsymbol{u}_\infty\|^{p'}_{\boldsymbol{L}^{p'}(\Omega)} \\ &\quad + \varepsilon\|\mathrm{curl}\,\boldsymbol{w}(t)\|^p_{\boldsymbol{L}^p(\Omega)}. \end{aligned}$$

Put

$$\begin{aligned} \xi(t) &= \|\mathrm{div}(\boldsymbol{f}(t)-\boldsymbol{f}_\infty)\|^{p'\wedge 2}_{L^{p'}(\Omega)} + \|\boldsymbol{f}(t)-\boldsymbol{f}_\infty\|^{p'\wedge 2}_{\boldsymbol{L}^{p'}(\Omega)} \\ &\quad + \|(\boldsymbol{h}(t)-\boldsymbol{h}_\infty)\times\boldsymbol{n}\|^{p'\wedge 2}_{\boldsymbol{W}^{-1/p',p'}(\Gamma)} + \|\mathrm{div}_\Gamma((\boldsymbol{h}(t)-\boldsymbol{h}_\infty)\times\boldsymbol{n})\|^{p'\wedge 2}_{W^{-1/p',p'}(\Gamma)}, \end{aligned}$$

where $p'\wedge 2 = \min\{p',2\}$, and

$$\zeta(t) = \|S_s^{(\infty)}(x,|\mathrm{curl}\,\boldsymbol{u}_\infty|^2)\mathrm{curl}\,\boldsymbol{u}_\infty - S_s(x,t,|\mathrm{curl}\,\boldsymbol{u}_\infty|^2)\mathrm{curl}\,\boldsymbol{u}_\infty\|^{p'}_{\boldsymbol{L}^{p'}(\Omega)}.$$

When $p\ge 2$, from Lemma 2.3, we have

$$\begin{aligned} &\int_\Omega (S_s(x,t,|\mathrm{curl}\,\boldsymbol{u}(t)|^2)\mathrm{curl}\,\boldsymbol{u}(t) - S_s(x,t,|\mathrm{curl}\,\boldsymbol{u}_\infty|^2)\mathrm{curl}\,\boldsymbol{u}_\infty)\cdot\mathrm{curl}\,\boldsymbol{w}(t)dx \\ &\qquad \ge \lambda\int_\Omega |\mathrm{curl}\,\boldsymbol{w}(t)|^p dx = \lambda\|\boldsymbol{w}(t)\|^p_{\mathbb{V}^p_T(\Omega)}. \end{aligned}$$

It follows from the Hölder inequality that

$$\int_\Omega |\boldsymbol{w}(t)|^2 dx \le C(p,\Omega)\left(\int_\Omega |\boldsymbol{w}(t)|^p dx\right)^{2/p},$$

so

$$\left(\int_\Omega |\boldsymbol{w}(t)|^2 dx\right)^{p/2} \le C\int_\Omega |\boldsymbol{w}(t)|^p dx \le C_2\|\boldsymbol{w}(t)\|^p_{\mathbb{V}^p_T(\Omega)}.$$

Hence, we have

$$\frac{1}{2}\frac{d}{dt}\int_\Omega |\boldsymbol{w}(t)|^2 dx + c\left(\int_\Omega |\boldsymbol{w}(t)|^2 dx\right)^{p/2} \le C\xi(t) + D\zeta(t). \qquad (5.7)$$

If we put

$$\phi(t) = \int_\Omega |\boldsymbol{w}(t)|^2 dx \text{ and } l(t) = 2C\xi(t) + 2D\zeta(t),$$

then we have

$$\phi'(t) + 2c\phi(t)^{p/2} \le l(t). \qquad (5.8)$$

5.1. The Degenerate Case $p > 2$

When $p > 2$, we use the following lemma due to Simon [12, p. 600].

Lemma 5.2. *Let $\phi(t)$ is a continuous positive function in an interval $I \subset \mathbb{R}$, and differentiable for a.e. $t \in I$ such that*

$$\phi'(t) + c(t)\phi(t)^{p/2} \le l(t) \text{ a.e. } t \in I,$$

where $p > 2$, $c(t) \ge 0$ and $l \in L^1(I)$. Then for any $t_0, t \in I$ with $t_0 \le t$,

$$\phi(t) \le \left(\frac{p-2}{2}\int_{t_0}^t c(\sigma)d\sigma\right)^{-2/(p-2)} + \int_{t_0}^t l(\sigma)d\sigma.$$

Applying this lemma with $t_0 = t/2$, $c(t) = 2c$ and $I = (0,\infty)$, we have the following.

Theorem 5.3. *When $p > 2$, if we assume that*

$$\int_{t/2}^t (\xi(\tau) + \zeta(\tau)d\tau \to 0 \text{ as } t \to \infty,$$

then we have

$$\|\boldsymbol{u}(t)-\boldsymbol{u}_\infty\|_{\boldsymbol{L}^2(\Omega)} \to 0 \text{ as } t \to \infty.$$

Furthermore, if $\boldsymbol{f} \in C(0,\infty;\boldsymbol{L}^{p'}(\Omega))$ *with* $\operatorname{div} \boldsymbol{f} \in C(0,\infty;L^{p'}(\Omega))$ *and* $\operatorname{div}_\Gamma(\boldsymbol{h}\times \boldsymbol{n}) \in C(0,\infty;W^{-1/p',p'}(\Gamma))$ *satisfy*

$$\|\boldsymbol{f}(t)-\boldsymbol{f}_\infty\|_{\boldsymbol{L}^{p'}(\Omega)} + \|\operatorname{div}(\boldsymbol{f}(t)-\boldsymbol{f}_\infty)\|_{L^{p'}(\Omega)} + \|\operatorname{div}_\Gamma((\boldsymbol{h}(t)-\boldsymbol{h}_\infty)\|_{W^{-1/p',p'}(\Gamma)} \to 0$$

as $t \to \infty$ *,then we have*

$$\|\pi(t)-\pi_\infty\|_{W^{1,p'}(\Omega)/\mathbb{R}} \to 0 \text{ as } t \to \infty.$$

5.2. The Case $p = 2$

From (5.7), we have

$$\frac{d}{dt}\int_\Omega |\boldsymbol{w}(t)|^2 dx + 2c\int_\Omega |\boldsymbol{w}(t)|^2 dx \le l(t) \le l_0, \tag{5.9}$$

where l_0 is a constant independent of t. We show that $\boldsymbol{w} \in L^\infty(0,\infty;\boldsymbol{L}^2(\Omega))$. Indeed, multiplying e^{2ct} to the above inequality, and then integrating over (σ,τ), we have

$$\int_\sigma^\tau \int_\Omega e^{2ct}\partial_t |\boldsymbol{w}(t)|^2 dxdt + 2c\int_\sigma^\tau e^{2ct}|\boldsymbol{w}(t)|^2 dxdt \le l_0 \int_\sigma^\tau e^{2ct}dt = \frac{l_0}{2c}(e^{2c\tau}-e^{2c\sigma}).$$

Since

$$\int_\sigma^\tau \int_\Omega e^{2ct}\partial_t |\boldsymbol{w}(t)|^2 dxdt = e^{2c\tau}\int_\Omega |\boldsymbol{w}(\tau)|^2 dx - e^{2c\sigma}\int_\Omega |\boldsymbol{w}(\sigma)|^2 dx \\ - 2c\int_\sigma^\tau \int_\Omega e^{2ct}|\boldsymbol{w}(t)|^2 dxdt,$$

we have

$$e^{2c\tau}\int_\Omega |\boldsymbol{w}(\tau)|^2 dx \le \frac{l_0}{2c}(e^{2c\tau}-e^{2c\sigma}) + e^{2c\sigma}\int_\Omega |\boldsymbol{w}(\sigma)|^2 dx.$$

If we put $\tau = t$ and $\sigma = 0$, we have

$$\int_\Omega |\boldsymbol{w}(t)|^2 dx \le \frac{l_0}{2c} + \int_\Omega |\boldsymbol{u}_0 - \boldsymbol{u}_\infty|^2 dx = l_1.$$

Here we use the following lemma (cf. Heraux [8, p. 286]).

Lemma 5.4. *Let $\phi(t)$ be a non-negative function, and absolutely continuous in any compact interval of $\mathbb{R}^+$ and $c > 0$ be a constant, and assume that $l(t)$ is a non-negative function and belongs to $L^1_{\text{loc}}(\mathbb{R}^+)$. If*

$$\phi'(t) + c\phi(t) \le l(t) \text{ for all } t \ge 0,$$

then for any $t_0, t \in \mathbb{R}^+$ with $t_0 \le t$,

$$\phi(t) \le e^{c(t_0-t)}\phi(t_0) + \frac{1}{1-e^{-c}} \sup_{\tau \ge t_0} \int_\tau^{\tau+1} l(\sigma)d\sigma.$$

Applying this lemma to (5.8) with

$$\phi(t) = \int_\Omega |\boldsymbol{w}(t)|^2 dx,$$

if we put t_0 fixed, then for any $t > t_0$, we have

$$\phi(t) \le e^{2c(t_0-t)}\phi(t_0) + \frac{1}{1-e^{-2c}} \sup_{\tau \ge t_0} \int_\tau^{\tau+1} l(\sigma)d\sigma.$$

Thus we have the following.

Theorem 5.5. *When $p = 2$, if*

$$\int_t^{t+1} (\xi(\tau) + \zeta(\tau))d\tau \to 0 \text{ as } t \to \infty,$$

then we have

$$\|\boldsymbol{u}(t) - \boldsymbol{u}_\infty\|_{L^2(\Omega)} \to 0 \text{ as } t \to \infty.$$

Furthermore, if $\boldsymbol{f} \in C(0,\infty; \boldsymbol{L}^2(\Omega))$ with $\operatorname{div} \boldsymbol{f} \in C(0,\infty; L^2(\Omega))$ and $\operatorname{div}_\Gamma(\boldsymbol{h} \times \boldsymbol{n}) \in C(0,\infty; H^{-1/2}(\Gamma))$ satisfy

$$\|\boldsymbol{f}(t) - \boldsymbol{f}_\infty\|_{L^2(\Omega)} + \|\operatorname{div}(\boldsymbol{f}(t) - \boldsymbol{f}_\infty)\|_{L^2(\Omega)} + \|\operatorname{div}_\Gamma((\boldsymbol{h}(t) - \boldsymbol{h}_\infty)\|_{H^{-1/2}(\Gamma)} \to 0$$

as $t \to \infty$, then we have

$$\|\pi(t) - \pi_\infty\|_{H^1(\Omega)/\mathbb{R}} \to 0 \text{ as } t \to \infty.$$

5.3. The Singular Case $6/5 \le p < 2$.

In this case, we assume that $6/5 \le p < 2$ and $S(x,t,s) = S^{(\infty)}(x,s)$. Assume that $\boldsymbol{f} \in L^\infty(0,\infty; \boldsymbol{L}^{p'}(\Omega)) \cap \boldsymbol{L}^2(\Omega_\infty)$ with $\operatorname{div} \boldsymbol{f} \in L^\infty(0,\infty; \boldsymbol{L}^{p'}(\Omega)) \cap \boldsymbol{L}^2(\Omega_\infty)$,

$$\boldsymbol{h} \times \boldsymbol{n} \in L^\infty(0,\infty; \boldsymbol{W}^{-1/p',p'}(\Gamma)) \cap W^{1,p'}(0,\infty; \boldsymbol{W}^{-1/p',p'}(\Gamma))$$

with

$$\operatorname{div}_\Gamma(\boldsymbol{h} \times \boldsymbol{n}) \in L^\infty(0,\infty; \boldsymbol{W}^{-1/p',p'}(\Gamma)) \cap L^2(0,\infty; H^{-1/2}(\Gamma))$$

and $\boldsymbol{u}_0 \in \widetilde{\mathbb{V}}^p_T(\Omega)$. Moreover, we assume that (4.2) holds. From (5.6) and Lemma 2.3, we have

$$\frac{d}{dt}\int_\Omega |\boldsymbol{w}(t)|^2 dx + \lambda \int_\Omega (|\operatorname{curl} \boldsymbol{u}(t)| + |\operatorname{curl} \boldsymbol{u}_\infty|)^{p-2} |\operatorname{curl} \boldsymbol{w}(t)|^2 dx \le I_1 + I_2.$$

We use the reverse Hölder inequality (cf. Sobolev [13, p. 8]) with $0 < s = p/2 < 1$ and $s' = p/(p-2) < 0$. We have

$$\begin{aligned}
&\int_{\widehat{\Omega}} (|\operatorname{curl} \boldsymbol{u}(t)| + |\operatorname{curl} \boldsymbol{u}_\infty|)^{p-2} |\operatorname{curl} \boldsymbol{w}(t)|^2 dx \\
&\qquad \ge \left(\int_{\widehat{\Omega}} |\operatorname{curl} \boldsymbol{w}(t)|^p dx \right)^{2/p} \left(\int_{\widehat{\Omega}} (|\operatorname{curl} \boldsymbol{u}(t)| + |\operatorname{curl} \boldsymbol{u}_\infty|)^p dx \right)^{(p-2)/2},
\end{aligned}$$

where

$$\widehat{\Omega} = \{x \in \Omega; ||\operatorname{curl} \boldsymbol{u}(x,t)| + |\operatorname{curl} \boldsymbol{u}_\infty(x)| \ne 0\}.$$

However, we can apply Proposition 4.3 to show

$$\int_\Omega (|\operatorname{curl} \boldsymbol{u}(t)| + |\operatorname{curl} \boldsymbol{u}_\infty|)^p dx \le C.$$

Thus we have

$$\int_\Omega (|\operatorname{curl} \boldsymbol{u}(t)| + |\operatorname{curl} \boldsymbol{u}_\infty|)^{p-2} |\operatorname{curl} \boldsymbol{w}(t)|^2 dx \ge c \left(\int_\Omega |\operatorname{curl} \boldsymbol{w}(t)|^p dx \right)^{2/p}.$$

Therefore there exists a constant $C > 0$ such that

$$\frac{1}{2}\frac{d}{dt}\int_\Omega |\boldsymbol{w}(t)|^2 dx + c \left(\int_\Omega |\operatorname{curl} \boldsymbol{w}(t)|^p dx \right)^{2/p} \le I_1 + I_2.$$

For any $\varepsilon > 0$, we have

$$\begin{aligned} |I_1| &\leq C(\varepsilon)\|\nabla(\pi(t)-\pi_\infty)\|^2_{\boldsymbol{L}^2(\Omega)} + \varepsilon\|\boldsymbol{w}(t)\|^2_{\boldsymbol{L}^2(\Omega)}, \\ |I_2| &\leq C(\varepsilon)\|\boldsymbol{f}(t)-\boldsymbol{f}_\infty\|^2_{\boldsymbol{L}^2(\Omega)} + \varepsilon\|\boldsymbol{w}(t)\|^2_{\boldsymbol{L}^2(\Omega)} \\ &\qquad + C(\varepsilon)\|(\boldsymbol{h}(t)-\boldsymbol{h}_\infty)\times\boldsymbol{n}\|^2_{\boldsymbol{W}^{-1/p',p'}(\Gamma)} + \varepsilon\|\boldsymbol{w}(t)\|^2_{\boldsymbol{W}^{1-1/p,p}(\Gamma)}. \end{aligned}$$

Since $6/5 \leq p < 2$, we have $\mathbb{V}^p_T(\Omega) \hookrightarrow \boldsymbol{L}^2(\Omega)$, so

$$\int_\Omega |\boldsymbol{w}(t)|^2 dx \leq C\left(\int_\Omega |\boldsymbol{w}(t)|^p dx\right)^{2/p} \leq C_1\left(\int_\Omega |\operatorname{curl}\boldsymbol{w}(t)|^p dx\right)^{2/p}.$$

If we choose $\varepsilon > 0$ small enough, then we have

$$\begin{aligned} \frac{1}{2}\frac{d}{dt}\int_\Omega |\boldsymbol{w}(t)|^2 dx + c\int_\Omega |\boldsymbol{w}(t)|^2 dx &\leq C(\|\nabla(\pi(t)-\pi_\infty)\|^2_{\boldsymbol{L}^2(\Omega)} \\ &\quad + \|\boldsymbol{f}(t)-\boldsymbol{f}_\infty\|^2_{\boldsymbol{L}^2(\Omega)} + \|(\boldsymbol{h}(t)-\boldsymbol{h}_\infty)\times\boldsymbol{n}\|_{\boldsymbol{W}^{-1/p',p'}(\Gamma)}). \end{aligned}$$

Since $p' > 2$, we have $L^{p'}(\Omega) \hookrightarrow L^2(\Omega)$. Taking (5.5) into consideration, we have

$$\begin{aligned} &\frac{1}{2}\frac{d}{dt}\int_\Omega |\boldsymbol{w}(t)|^2 dx + c\int_\Omega |\boldsymbol{w}(t)|^2 dx \\ &\leq C(\|\boldsymbol{f}(t)-\boldsymbol{f}_\infty\|^2_{\boldsymbol{L}^{p'}(\Omega)} + \|\operatorname{div}(\boldsymbol{f}(t)-\boldsymbol{f}_\infty)\|^2_{L^{p'}(\Omega)} + \|(\boldsymbol{h}(t)-\boldsymbol{h}_\infty)\times\boldsymbol{n}\|^2_{\boldsymbol{W}^{-1/p',p'}(\Gamma)} \\ &\qquad + \|\operatorname{div}_\Gamma(\boldsymbol{h}(t)-\boldsymbol{h}_\infty)\times\boldsymbol{n}\|^2_{W^{-1/p',p'}(\Gamma)}) = Cl(t). \end{aligned}$$

By the similar method of the case $p = 2$, we get the following.

Theorem 5.6. *Assume that $S(x,t,s) = S^{(\infty)}(x,s)$ and $6/5 \leq p < 2$. Furthermore, we assume that $\boldsymbol{f}, \boldsymbol{h}$ and $\boldsymbol{u}_0$ satisfy the assumptions of Proposition 4.3 with $T = \infty$. If*

$$\int_t^{t+1} l(\tau)d\tau \to 0 \text{ as } t \to \infty,$$

then we have

$$\|\boldsymbol{u}(t)-\boldsymbol{u}_\infty\|_{L^2(\Omega)} \to 0 \text{ as } t \to \infty.$$

Furthermore, if $\boldsymbol{f} \in C(0,\infty;\boldsymbol{L}^{p'}(\Omega))$ with $\operatorname{div}\boldsymbol{f} \in C(0,\infty;L^{p'}(\Omega))$ and $\operatorname{div}_\Gamma(\boldsymbol{h}\times\boldsymbol{n}) \in C(0,\infty;W^{-1/p',p'}(\Gamma))$ satisfy

$$\|\boldsymbol{f}(t)-\boldsymbol{f}_\infty\|_{\boldsymbol{L}^{p'}(\Omega)} + \|\operatorname{div}(\boldsymbol{f}(t)-\boldsymbol{f}_\infty)\|_{L^{p'}(\Omega)} + \|\operatorname{div}_\Gamma((\boldsymbol{h}(t)-\boldsymbol{h}_\infty)\|_{W^{-1/p',p'}(\Gamma)} \to 0$$

as $t \to \infty$, then we have

$$\|\pi(t)-\pi_\infty\|_{W^{1,p'}(\Omega)/\mathbb{R}} \to 0 \text{ as } t \to \infty.$$

Appendix. Proof of Lemma 3.4

In this appendix, we prove Lemma 3.4. Let $\boldsymbol{z} \in \boldsymbol{L}^{p'}(\Omega)$ with $\operatorname{curl} \boldsymbol{z} \in \mathbb{X}_T^p(\Omega)' \subset \boldsymbol{W}^{-1,p'}(\Omega)$. For any $\boldsymbol{\varphi} \in \boldsymbol{W}^{1-1/p,p}(\Gamma)$ with $\boldsymbol{\varphi} \times \boldsymbol{n} = \boldsymbol{0}$ on Γ, there exists $\widetilde{\boldsymbol{\varphi}} \in \boldsymbol{W}^{1,p}(\Omega)$ such that $\widetilde{\boldsymbol{\varphi}} = \boldsymbol{\varphi}$ on Γ and satisfies

$$\|\widetilde{\boldsymbol{\varphi}}\|_{\boldsymbol{W}^{1,p}(\Omega)} \le C\|\boldsymbol{\varphi}\|_{\boldsymbol{W}^{1-1/p,p}(\Gamma)}.$$

We note that $\widetilde{\boldsymbol{\varphi}} \in \mathbb{X}_T^p(\Omega)$. Define

$$\langle \boldsymbol{z} \times \boldsymbol{n}, \boldsymbol{\varphi}\rangle = \langle \operatorname{curl} \boldsymbol{z}, \widetilde{\boldsymbol{\varphi}}\rangle_{\mathbb{X}_T^p(\Omega)', \mathbb{X}_T^p(\Omega)} - \int_\Omega \boldsymbol{z} \cdot \operatorname{curl} \widetilde{\boldsymbol{\varphi}} dx. \tag{5.10}$$

We note that if $\boldsymbol{z} \in \boldsymbol{C}^1(\overline{\Omega})$, then we have

$$\langle \operatorname{curl} \boldsymbol{z}, \widetilde{\boldsymbol{\varphi}}\rangle_{\widetilde{\mathbb{X}}_T^p(\Omega)', \widetilde{\mathbb{X}}_T^p(\Omega)} = \int_\Omega \operatorname{curl} \boldsymbol{z} \cdot \widetilde{\boldsymbol{\varphi}} dx.$$

For $\boldsymbol{z} \in \boldsymbol{L}^{p'}(\Omega)$ with $\operatorname{curl} \boldsymbol{z} \in \mathbb{X}_T^p(\Omega)'$, we can choose $\boldsymbol{z}_j \in \boldsymbol{C}^1(\overline{\Omega})$ such that $\boldsymbol{z}_j \to \boldsymbol{z}$ in $\boldsymbol{L}^{p'}(\Omega)$. Thus, if $\boldsymbol{\varphi} \in \boldsymbol{W}_0^{1,p}(\Omega)$,

$$\langle \operatorname{curl} \boldsymbol{z}_j, \boldsymbol{\varphi}\rangle_{\mathbb{X}_T^p(\Omega)', \mathbb{X}_T^p(\Omega)} = \int_\Omega \boldsymbol{z}_j \cdot \operatorname{curl} \boldsymbol{\varphi} dx \to \int_\Omega \boldsymbol{z} \cdot \operatorname{curl} \boldsymbol{\varphi} dx$$

as $j \to \infty$. On the other hand, since $\operatorname{curl} \boldsymbol{z}_j \to \operatorname{curl} \boldsymbol{z}$ in $\boldsymbol{W}^{-1,p'}(\Omega)$, we have

$$\langle \operatorname{curl} \boldsymbol{z}_j, \boldsymbol{\varphi}\rangle_{\mathbb{X}_T^p(\Omega)', \mathbb{X}_T^p(\Omega)} = \langle \operatorname{curl} \boldsymbol{z}_j, \boldsymbol{\varphi}\rangle_{\boldsymbol{W}^{-1,p'}(\Omega), \boldsymbol{W}_0^{1,p}(\Omega)} \to \langle \operatorname{curl} \boldsymbol{z}, \boldsymbol{\varphi}\rangle_{\mathbb{X}_T^p(\Omega)', \mathbb{X}_T^p(\Omega)}$$

as $j \to \infty$. Hence

$$\langle \operatorname{curl} \boldsymbol{z}, \boldsymbol{\varphi}\rangle_{\mathbb{X}_T^p(\Omega)', \mathbb{X}_T^p(\Omega)} = \int_\Omega \boldsymbol{z} \cdot \operatorname{curl} \boldsymbol{\varphi} dx \text{ for all } \boldsymbol{\varphi} \in \boldsymbol{W}_0^{1,p}(\Omega).$$

This implies that (5.10) is independent of the choice of $\widetilde{\boldsymbol{\varphi}}$ such that $\widetilde{\boldsymbol{\varphi}} = \boldsymbol{\varphi}$ on Γ. Since

$$\begin{aligned}|\langle \boldsymbol{z} \times \boldsymbol{n}, \boldsymbol{\varphi}\rangle| &\le C(\|\operatorname{curl} \boldsymbol{z}\|_{\widetilde{\mathbb{X}}_T^p(\Omega)'} + \|\boldsymbol{z}\|_{\boldsymbol{L}^{p'}(\Omega)})\|\widetilde{\boldsymbol{\varphi}}\|_{\boldsymbol{W}^{1,p}(\Omega)} \\ &\le C_1(\|\operatorname{curl} \boldsymbol{z}\|_{\widetilde{\mathbb{X}}_T^p(\Omega)'} + \|\boldsymbol{z}\|_{\boldsymbol{L}^{p'}(\Omega)})\|\boldsymbol{\varphi}\|_{\boldsymbol{W}^{1-1/p,p}(\Gamma)}.\end{aligned}$$

Thus $\boldsymbol{z} \times \boldsymbol{n} \in \boldsymbol{W}^{-1/p',p'}(\Gamma)$ and (3.11) holds.

REFERENCES

[1] Amrouche C., Seloula N. H., L^p-theory for vector potentials and Sobolev's inequality for vector fields. Application to the Stokes equations with pressure boundary conditions, *Math. Models and Methods in Appl. Sci.*, 23 (2013) 37-92.

[2] Amrouche C., Seloula N. H., L^p-theory for vector potentials and Sobolev's inequality for vector fields., *C. R. Acad. Sci. Paris, Ser. 1.*, 349 (2011) 529-534.

[3] Aramaki J., Existence and regularity for the Neumann problem to teh Poisson equation and an application to the Maxwell-Stokes type equation, *Comm. Math. Anal.* 21(1), (2018) 54-66.

[4] Aramaki J., *Existence and regularity of a weak solution to a class of systems in a multi-connected domain*, submitted for publication.

[5] Aramaki J., *Necessary and sufficient conditions for the existence of a weak solution to the Maxwell-Stokes equation*, submitted for publication.

[6] Dautray R. and Lions J. L., *Mathematical Analysis and Numerical Method for Science and Technology* Vol. 3, Springer Verlag, New York, (1990).

[7] Girault V. and Raviart P. A., *Finite Element Methods for Navier-Stokes equations*, Springer-Verlag, Berlin, Heidelberg, New York, Tokyo, (1979).

[8] Haraux A., Nonlinear Evolution Equations-Global Behavior of Solutions, *Lecture Notes in Math.*, 841, Springer-Verlag, Berlin-New York, (1981).

[9] Lions J. L., *Quelque Méthodes de Résolution des Problémes aux Limites Non Linéaires* [*Some Methods of Solving Problems with Non Linear Boundaries*], Dunod Gauthier-Villars, (1969).

[10] Miranda F., Rodrigues J. F. and Santos L., On a p-curl system arising in electromagnetism, *Discrete and Cont. Dyn. Systems Ser. S*, 6(3), (2012), 605-629.

[11] Mitrea D., Mitrea M. and Pipher J, Vector potential theory on nonsmooth domains in $\mathbb{R}^3$ and applications to electromagnetic scattering, *Fourier J., Anal. Appl.* **3** no. 2 (1997), 131-192.

[12] Simon J., Quelque propriétés de solutions d'équations et d'évolution paraboliques non linéires, *Ann. Scuola Norm. Pisa Cl. Sci.* (4), 2, (1975), 585-609.

[13] Sobolev S., Applications of Functional Analysis in Mathematical Physics, *Translation of Mathematical Monographs*, Vol. 7, AMS, Providence, R. I., (1963).

[14] Yin H, Li B. Q. and Zou J., A degenerate evolution system modeling Bean's critical state type-II superconductors, *Discrete and Continuous Dynamical Systems*, 8, (2002) 781-794.

[15] Zheng S., *Nonlinear Evolution Equations*, Chapman and Hall/CRC, (2004).

Related Nova Publications

From Maxwell's Equations to Free and Guided Electromagnetic Waves: An Introduction for First-Year Undergraduates

Manuel Quesada-Pérez
and José Alberto Maroto-Centeno
University of Jaén, Department of Physics,
Escuela Politécnica Superior de Linares, Linares, Jaén, Spain

Abstract

Maxwell's equations and the discovery of electromagnetic waves changed the world. Can you imagine how our everyday life would be without telephone, radio, television, mobile phones and internet? It is thanks to Maxwell's equations that we understand what electromagnetic waves are and how they are generated, propagated and detected. These equations can even change our perception of nature when they are really understood, but their power and elegance is completely appreciated when they are expressed in differential form. Moreover, this form is extremely useful dealing with some issues, such as the propagation of electromagnetic waves.

This book can be very useful for undergraduates that must face the differential form of Maxwell's equations and its application to electromagnetic waves for the first time because:

- Maxwell's equations and the propagation of free and guided waves are introduced with mathematical rigor but without requiring previous knowledge on advanced vector calculus or theory of functions of complex variable. Only elementary notions of calculus, vectors, geometry and trigonometry are required.
- The definition and notation of vector operators (divergence and curl) are explained in easy mathematical terms.
- An assortment of examples and solved problems illustrates how Maxwell's equations work in differential form and electromagnetic waves propagate in free space and confined media.
- Several examples of waveguides are analyzed with the help of an intuitive picture and without solving coupled partial differential equations to elucidate the peculiarities of guided propagation. Advanced mathematical treatments can sometimes hide the physical understanding of phenomena.

In short, this book will help students to bridge the gap between elementary and advances treatments on Maxwell's equation and electromagnetic waves. Thus, it can be a good training for electromagnetism courses.

Maxwell's Electromagnetic Equations, Elementary Introduction[*]

George J. Spix, and V. M. Red'kov[†]
Institute of Physics, National Academy of Sciences of Belarus

There is a need to present the detail mathematical steps that are requiredto prove the equations of Maxwell Text books and course instruction do givestudents a firm grasp of the equations and their applications. What is oftenmissing is the step by step exposition leading from the basic experiments to theestablished equations. The following book fills this gap admirably. These arepresented for students and erstwhile students who are interested in how the greatphysicists derived and proved their equations.

All of the math and physics presented here are more than covered in the fouryear college engineering course. However, students with a passing interest inhigh school math can easily understand all that is contained in these three papers. For some readers, there will be the revisiting of old friends. Other readerswill get the "Ah, Ha!" experience. The examples in using the equations arepresented in great detail and are easy to follow.

---

[*] The full version of this chapter can be found in *Systemic, Cellular and Molecular Mechanisms of Physiological Functions and Their Disorders (Proceedings of I. Beritashvili Center for Experimental Biomedicine – 2018)*, edited by Professor Nodar P. Mitagvaria and Professor Nargiz G. Nachkebia, published by Nova Science Publishers, Inc, New York, 2018.

[†] Corresponding Author's E-mail: v.redkov@ifanbel.bas-net.by.

BIBLIOGRAPHY

An Introduction to Mathematical Modeling: A Course in Mechanics

LCCN	2011012204
Type of material	Book
Personal name	Oden, J. Tinsley (John Tinsley), 1936-
Main title	An introduction to mathematical modeling: A course in mechanics / J. Tinsley Oden.
Published/Created	Hoboken, N.J.: Wiley, c2011.
Description	xiv, 334 p.; 24 cm.
ISBN	9781118019030 (hardback)
LC classification	QA807.O34 2011
Summary	"An important resource, this book provides a short-course in nonlinear continuum mechanics, contains a brief account of electromagnetic wave theory and Maxwell's equations as well as an introductory account of quantum mechanics, and presents a brief introduction to statistical mechanics of systems in thermodynamic equilibrium. Also included is information on continuum mechanics, electrodynamics, quantum mechanics, and statistical mechanics. The author approaches mechanics as the branch of physics and

mathematical science concerned with describing the motion of bodies, including their deformation and temperature changes, under the action of forces, and if the study of the propagation of waves and the transformation of energy in physical systems are added, then the term mechanics does indeed apply to everything that is covered within the book. "-- Provided by publisher.

"An important resource, this book provides a short-course in nonlinear continuum mechanics, contains a brief account of electromagnetic wave theory and Maxwell's equations as well as an introductory account of quantum mechanics, and presents a brief introduction to statistical mechanics of systems in thermodynamic equilibrium. Also included is information on continuum mechanics, electrodynamics, quantum mechanics, and statistical mechanics"-- Provided by publisher.

Contents Preface. I. Nonlinear Continuum Mechanics. 1. Kinematics of Deformable Bodies. 2. Mass and Momentum. 3. Force and Stress in Deformable Bodies. 4. The Principles of Balance of Linear and Angular Momentum. 5. The Principle of Conservation of Energy. 6. Thermodynamics of Continua and the Second Law. 7. Constitutive Equations. 8. Examples and Applications. II. Electromagnetic Field Theory and Quantum Mechanics. 9. Electromagnetic Waves. 10. Introduction to Quantum Mechanics. 11. Dynamical Variables and Observables in Quantum Mechanics: The Mathematical Formalism. 12. Applications: The Harmonic Oscillator and the Hydrogen Atom. 13. Spin and Pauli's Principle. 14. Atomic and Molecular

	Structure. 15. Ab Initio Methods: Approximate Methods and Density Functional Theory. III. Statistical Mechanics. 16. Basic Concepts: Ensembles, Distribution Functions and Averages. 17. Statistical Mechanics Basis of Classical Thermodynamics. Exercises. References.
Subjects	Mechanics, Analytic. Mathematics / General.
Notes	Includes bibliographical references (p. 317-323) and index.
Series	Wiley series in computational mechanics

Analysis and Modeling of Radio Wave Propagation

LCCN	2016045806
Type of material	Book
Personal name	Coleman, Christopher, 1950- author.
Main title	Analysis and modeling of radio wave propagation / Christopher John Coleman, University of Adelaide.
Published/Produced	Cambridge, United Kingdom; New York, NY: Cambridge University Press, [2017] ©2016
Links	Contributor biographical information https://www.loc.gov/catdir/enhancements/fy1701/2016045806-b.html Publisher description https://www.loc.gov/catdir/enhancements/fy1701/2016045806-d.html Table of contents only https://www.loc.gov/catdir/enhancements/fy1701/2016045806-t.html
ISBN	9781107175563 (hardback; alk. paper) 1107175569 (hardback; alk. paper)
LC classification	TK6553.C635 2017
Summary	"This comprehensive guide helps readers understand the theory and techniques needed to

analyze and model radio wave propagation in complex environments. All of the essential topics are covered, from the fundamental concepts of radio systems, to complex propagation phenomena. These topics include diffraction, ray tracing, scattering, atmospheric ducting, ionospheric ducting, scintillation, and propagation through both urban and non-urban environments. Emphasis is placed on practical procedures, with detailed discussion of numerical and mathematical methods providing readers with the necessary skills to build their own propagation models and develop their own techniques. MATLAB functions illustrating key modeling ideas are provided online. This is an invaluable resource for anyone wanting to use propagation models to understand the performance of radio systems for navigation, radar, communications, or broadcasting"-- Provided by publisher.

Contents	Basic concepts page -- The fundamentals of electromagnetic waves -- The reciprocity, compensation and extinction theorems -- The effect of obstructions upon radio wave propagation -- Geometric optics -- Propagation through irregular media -- The approximate solution of Maxwell's equations -- Propagation in the ionospheric duct -- Propagation in the lower atmosphere -- Transionospheric propagation and scintillation.
Subjects	Radio wave propagation. Radio wave propagation--Mathematical models. Electromagnetic waves.
Notes	Includes bibliographical references and index.

Biomedical Applications of Light Scattering

LCCN	2009036813
Type of material	Book
Personal name	Wax, Adam.
Main title	Biomedical applications of light scattering / Adam Wax, Vadim Backman.
Published/Created	New York: McGraw-Hill, c2010.
Description	xv, 368 p., [16] p. of plates: ill. (some col.); 24 cm.
ISBN	9780071598804 (alk. paper) 0071598804 (alk. paper)
LC classification	QP82.2.L5 W39 2010
Related names	Backman, Vadim.
Contents	Classical light scattering models -- Light scattering from continuous random media -- Modeling of light scattering by biological tissues via computational solution of Maxwell's equations -- Interferometric synthetic aperture microscopy -- Light scattering as a tool in cell biology -- Light absorption and scattering spectroscopic microscopies -- Light scattering in confocal reflectance microscopy -- Tissue ultrastructure scattering with near-infrared spectroscopy: ex vivo and in vivo interpretation -- Dynamic light scattering and motility-contrast imaging of living tissue -- Laser speckle contrast imaging of blood flow -- Elastic-scattering spectroscopy for optical biopsy: probe designs and analytical methods for clinical applications -- Differential pathlength spectroscopy -- Angle-resolved low-coherence interferometry: depth-resolved light scattering for detecting neoplasia -- Enhanced backscattering and low-coherence enhanced backscattering spectroscopy.

Subjects	Light--Scattering. Microscopy. Spectrum analysis. Light. Scattering, Radiation. Biomedical Technology.
Notes	Includes bibliographical references and index.
Series	Biophotonics series McGraw-Hill biophotonics.

Classical and Quantum Thermal Physics

LCCN	2016030572
Type of material	Book
Personal name	Prasad, R. (Emeritus Professor of Physics), author.
Main title	Classical and quantum thermal physics / R. Prasad.
Published/Produced	Daryaganj, Delhi, India: Cambridge University Press, [2016] ©2016
Description	xxii, 575 pages; 25 cm
Links	Contributor biographical information https://www.loc.gov/catdir/enhancements/fy1618/2016030572-b.html Publisher description https://www.loc.gov/catdir/enhancements/fy1618/2016030572-d.html Table of contents only https://www.loc.gov/catdir/enhancements/fy1618/2016030572-t.html
ISBN	9781107172883 (hardback; alk. paper) 1107172888 (hardback; alk. paper)
LC classification	QC311.P78 2016
Summary	"Discusses the interactions of heat energy and matter"-- Provided by publisher.
Contents	The kinetic theory of gases -- Ideal to a real gas, viscosity, conductivity and diffusion --

	Thermodynamics: definitions and the Zeroth law -- First law of thermodynamics and some of its applications -- Second law of thermodynamics and some of its applications -- Tds equations and their applications -- Thermodynamic functions, potentials, Maxwell's equations, the third law and equilibrium -- Some applications of thermodynamics to problems of physics and engineering -- Application of thermodynamics to chemical reactions -- Quantum thermodynamics -- Some applications of quantum thermodynamics -- Introduction to the thermodynamics of irreversible processes.
Subjects	Thermodynamics. Quantum theory. Kinetic theory of gases.
Notes	Includes bibliographical references and index.

Classical field theory: on electrodynamics, non-abelian gauge theories and gravitation

LCCN	2012937327
Type of material	Book
Personal name	Scheck, Florian, 1936-
Main title	Classical field theory: on electrodynamics, non-abelian gauge theories and gravitation / Florian Scheck.
Published/Created	Berlin; New York: Springer, c2012.
Description	xii, 433 p.: ports.; 24 cm.
ISBN	9783642279843 (hdbk.: acid-free paper) 3642279848 (hdbk.: acid-free paper) 9783642279850 (electronic bk.) 3642279856 (electronic bk.)
LC classification	QC173.7.S34 2012
Contents	Maxwell's Equations -- Symmetries and

	Covariance of the Maxwell Equations -- Maxwell Theory as a Classical Field Theory -- Simple Applications of Maxwell Theory -- Local Gauge Theories -- Classical Field Theory of Gravitation.
Subjects	Field theory (Physics) Electrodynamics. Gauge fields (Physics)
Notes	Includes bibliographical references and index.
Series	Graduate texts in physics, 1868-4513 Graduate texts in physics.

Classical Mechanics and Electrodynamics

LCCN	2018049591
Type of material	Book
Personal name	Leinaas, Jon Magne, 1946- author.
Main title	Classical mechanics and electrodynamics / Jon Magne Leinaas (University of Oslo, Norway).
Published/Produced	Singapore; Hackensack, NJ: World Scientific Publishing Co. Pte. Ltd., [2019] ©2019
Description	xii, 350 pages; 23 cm
ISBN	9789813279360 (hardcover; alk. paper) 9813279362 (hardcover; alk. paper)
LC classification	QA805.L43 2019
Summary	"The book gives a general introduction to classical theoretical physics, in the fields of mechanics, relativity and electromagnetism. It is analytical in approach and detailed in the derivations of physical consequences from the fundamental principles in each of the fields. The book is aimed at physics students in the last year of their undergraduate or first year of their graduate studies"-- Provided by publisher.

Contents	Generalized coordinates -- Lagrange's equations -- Hamiltonian dynamics -- The four-dimensional space-time -- Consequences of the Lorentz transformations -- Four-vector formalism and covariant equations -- Relativistic kinematics -- Relativistic dynamics -- Maxwell's equations -- Electromagnetic field dynamics -- Maxwell's equations with stationary sources -- Electromagnetic radiation.
Subjects	Mechanics, Analytic. Electrodynamics. Mathematical physics.
Notes	Includes bibliographical references and index.

Computational Methods for Electromagnetic and Optical Systems

LCCN	2010045338
Type of material	Book
Personal name	Jarem, John M., 1948-
Main title	Computational methods for electromagnetic and optical systems / John M. Jarem, Partha P. Banerjee.
Edition	2nd ed.
Published/Created	Boca Raton, FL: CRC Press, c2011.
Description	xv, 416 p.: ill.; 27 cm.
ISBN	9781439804223 (hardback) 1439804222 (hardback)
LC classification	QC760.J47 2011
Related names	Banerjee, Partha P.
Summary	"This text introduces and examines a variety of spectral computational techniques - including k-space theory, Floquet theory and beam propagation - that are used to analyze electromagnetic and optical problems. The book also presents a solution to Maxwell's equations

from a set of first order coupled partial differential equations"--Provided by publisher.

Contents 1.1. Introduction -- 1.2. Fourier Series and Its Properties -- 1.3. Fourier Transform -- 1.4. Hankel Transform -- 1.5. Discrete Fourier Transform -- 1.6. Review of Eigenanalysis -- Problems -- References -- 2.1. Introduction -- 2.2. Transfer Function for Propagation -- 2.3. Split-Step Beam Propagation Method -- 2.4. Beam Propagation in Linear Media -- 2.4.1. Linear Free-Space Beam Propagation -- 2.4.2. Propagation of Gaussian Beam through Graded Index Medium -- 2.5. Beam Propagation through Diffraction Gratings: Acoustooptic Diffraction -- 2.6. Beam Propagation in Kerr-Type Nonlinear Media -- 2.6.1. Nonlinear Schrodinger Equation -- 2.6.2. Simulation of Self-Focusing Using Adaptive Fourier and Fourier-Hankel Transform Methods -- 2.7. Beam Propagation and Coupling in Photorefractive Media -- 2.7.1. Basic Photorefractive Physics -- 2.7.2. Induced Transmission Gratings -- 2.7.3. Induced Reflection Gratings and Bidirectional Beam Propagation Method -- 2.8. z-Scan Method -- 2.8.1. Model for Beam Propagation through PR Lithium Niobate -- 2.8.2. z-Scan: Analytical Results, Simulations, and Sample Experiments -- Problems -- References -- 3.1. Introduction -- 3.2. Maxwell's Equations -- 3.3. Constitutive Relations: Frequency Dependence and Chirality -- 3.3.1. Constitutive Relations and Frequency Dependence -- 3.3.2. Constitutive Relations for Chiral Media -- 3.4. Plane Wave Propagation through Linear Homogeneous Isotropic Media -- 3.4.1. Dispersive Media -- 3.4.2. Chiral Media --

3.5. Power Flow, Stored Energy, Energy Velocity, Group Velocity, and Phase Velocity -- 3.6. Metamaterials and Negative Index Media -- 3.6.1. Beam Propagation in NIMs -- 3.7. Propagation through Photonic Band Gap Structures: The Transfer Matrix Method -- 3.7.1. Periodic PIM-NIM Structures -- 3.7.2. EM Propagation in Complex Structures -- Problems -- References -- 4.1. Introduction -- 4.2. State Variable Analysis of an Isotropic Layer -- 4.2.1. Introduction -- 4.2.2. Analysis -- 4.2.3. Complex Poynting Theorem -- 4.2.4. State Variable Analysis of an Isotropic Layer in Free Space -- 4.2.5. State Variable Analysis of a Radar Absorbing Layer -- 4.2.6. State Variable Analysis of a Source in Isotropic Layered Media -- 4.3. State Variable Analysis of an Anisotropic Layer -- 4.3.1. Introduction -- 4.3.2. Basic Equations -- 4.3.3. Numerical Results -- 4.4. One-Dimensional k-Space State Variable Solution -- 4.4.1. Introduction -- 4.4.2. k-Space Formulation -- 4.4.3. Ground Plane Slot Waveguide System -- 4.4.4. Ground Plane Slot Waveguide System, Numerical Results -- Problems -- References -- 5.1. Introduction -- 5.2. H-Mode Planar Diffraction Grating Analysis -- 5.2.1. Full-Field Formulation -- 5.2.2. Differential Equation Method -- 5.2.3. Numerical Results -- 5.2.4. Diffraction Grating Mirror -- 5.3. Application of RCWA and the Complex Poynting Theorem to E-Mode Planar Diffraction Grating Analysis -- 5.3.1. E-Mode RCWA Formulation -- 5.3.2. Complex Poynting Theorem -- 5.3.2.1. Sample Calculation of PuWE -- 5.3.2.2. Other Poynting Theorem Integrals -- 5.3.2.3.

Simplification of Results and Normalization -- 5.3.3. Numerical Results -- 5.4. Multilayer Analysis of E-Mode Diffraction Gratings -- 5.4.1. E-Mode Formulation -- 5.4.2. Numerical Results -- 5.5. Crossed Diffraction Grating -- 5.5.1. Crossed Diffraction Grating Formulation -- 5.5.2. Numerical Results -- Problems -- References -- 6.1. Introduction to Photorefractive Materials -- 6.2. Dynamic Nonlinear Model for Diffusion-Controlled PR Materials -- 6.3. Approximate Analysis -- 6.3.1. Numerical Algorithm -- 6.3.2. TE Numerical Simulation Results -- 6.3.3. TM Numerical Simulation Results -- 6.3.4. Discussion of Results from Approximate Analysis -- 6.4. Exact Analysis -- 6.4.1. Finite Difference Kukhtarev Analysis -- 6.4.2. TM Numerical Simulation Results -- 6.5. Reflection Gratings -- 6.5.1. RCWA Optical Field Analysis -- 6.5.2. Material Analysis -- 6.5.3. Numerical Results -- 6.6. Conclusion -- Problems -- References -- 7.1. Introduction -- 7.2. Rigorous Coupled Wave Analysis Circular Cylindrical Systems -- 7.3. Rigorous Coupled Wave Analysis Mathematical Formulation -- 7.3.1. Introduction -- 7.3.2. Basic Equations -- 7.3.3. Numerical Results -- 7.4. Anisotropic Cylindrical Scattering -- 7.4.1. Introduction -- 7.4.2. State Variable Analysis -- 7.4.3. Numerical Results -- 7.5. Spherical Inhomogeneous Analysis -- 7.5.1. Introduction -- 7.5.2. Rigorous Coupled Wave Theory Formulation -- 7.5.3. Numerical Results -- Problems -- References -- 8.1. Introduction -- 8.2. RCWA Bipolar Coordinate Formulation -- 8.2.1. Bipolar and Eccentric Circular Cylindrical,

Scattering Region Coordinate Description -- 8.2.2. Bipolar RCWA State Variable Formulation -- 8.2.3. Second-Order Differential Matrix Formulation -- 8.2.4. Thin-Layer, Bipolar Coordinate Eigenfunction Solution -- 8.3. Bessel Function Solutions in Homogeneous Regions of Scattering System -- 8.4. Thin-Layer SV Solution in the Inhomogeneous Region of the Scattering System -- 8.5. Matching of EM Boundary Conditions at Interior-Exterior Interfaces of the Scattering System -- 8.5.1. Bipolar and Circular Cylindrical Coordinate Relations -- 8.5.2. Details of Region 2 (Inhomogenous Region) Region 3 (Homogenous Interior Region) EM Boundary Value Matching -- 8.5.3. Region 0 (Homogenous Exterior Region) Region 2 (Inhomogenous Region) EM Boundary Value Matching -- 8.5.4. Details of Layer-to-Layer EM Boundary Value Matching in the Inhomogeneous Region -- 8.5.5. Inhomogeneous Region Ladder-Matrix -- 8.6. Region 1 Region 3 Bessel-Fourier Coefficient Transfer Matrix -- 8.7. Overall System Matrix -- 8.8. Alternate Forms of the Bessel-Fourier Coefficient Transfer Matrix -- 8.9. Bistatic Scattering Width -- 8.10. Validation of Numerical Results -- 8.11. Numerical Results, Examples of Scattering from Homogeneous and Inhomogeneous Material Objects -- 8.12. Error and Convergence Analysis -- 8.13. Summary, Conclusions, and Future Work -- Problems -- Appendix 8. A -- Appendix 8. B -- References -- 9.1. Introduction -- 9.2. Case Study I: Fourier Series Expansion, Eigenvalue and Eigenfunction Analysis, and Transfer Matrix Analysis -- 9.3.

Case Study II: Comparison of KPE BA, BC Validation Methods, and SV Methods for Relatively Small Diameter Scattering Objects -- 9.4. Case Study III: Comparison of BA, BC, and SV Methods for Gradually, Stepped-Up, Index Profile Scattering Objects -- 9.5. Case Study IV: Comparison of BA, BC, and SV Methods for Mismatched, Index Profile, Scattering Objects -- 9.6. Case Study V: Comparison of BA, BC, and SV Methods for Gradually, Stepped-Up, Index Scattering Objects with High Index Core -- 9.7. Case Study VI: Calculation and Convergence Analysis of EM Fields of an Inhomogeneous Region Material Object Using the SV Method, Δ epsilon = 1, α = 5.5, Λ = 0, Example -- 9.8. Case Study VII: Calculation and Convergence Analysis of EM Fields of an Inhomogeneous Region Material Object Using the SV Method, Δ epsilon = 0.4, α = 5.5, Λ = 0 Example -- 9.9. Case Study VIII: Comparison of Homogeneous and Inhomogeneous Region Bistatic Line Widths -- 9.10. Case Study IX: Conservation of Power Analysis -- Appendix 9. A: Interpolation Equations.

Subjects Electromagnetism--Mathematics.
Electromagnetism--Industrial applications.
Optics--Mathematics.
Optics--Industrial applications.

Notes Includes bibliographical references and index.

Series Optical science and engineering; 149
Optical science and engineering (Boca Raton, Fla.); 149.

Computing the Flow of Light: Nonstandard FDTD Methodologies for Photonics Design

LCCN	2016033109
Type of material	Book
Personal name	Cole, James B. (James Bradford), author.
Main title	Computing the flow of light: nonstandard FDTD methodologies for photonics design / James B. Cole, Saswatee Banerjee.
Published/Produced	Bellingham, Washington, USA: SPIE Press, [2017] ©2017
Description	xvi, 413 pages: illustrations (some color); 26 cm + 1 CD-ROM (4 3/4 in.)
ISBN	9781510604810 (softcover) 1510604812 (softcover) (pdf) (pdf) (epub) (epub) (Kindle/mobi) (Kindle/mobi)
LC classification	TA1522.C65 2017
Related names	Banerjee, Saswatee, author.
Summary	"Finite difference time domain (FDTD) computes the time evolution of a system at discrete time steps, and the resulting periodic visualization yields insight into the system. FDTD and FDTD-like methods can be used to solve a wide variety of problems, including--but not limited to--the wave equation, Maxwell's equations, and the Schrödinger equation. In addition to introducing useful new methodologies, this book provides readers with analytical background and simulation examples that will help them develop their own

	methodologies to solve yet-to-be-posed problems. The book is written for students, engineers, and researchers grappling with problems that cannot be solved analytically. It could also be used as a textbook for a mathematical physics or engineering class"-- Provided by publisher.
Contents	Finite difference approximations -- Accuracy, stability and convergence of numerical algorithms -- Introduction -- Finite difference models of the simple harmonic oscillator -- The one-dimensional wave equation -- Finite difference time domain algorithms for the one-dimensional wave equation -- Program development and applications of finite difference time domain algorithms in one-dimension -- Finite difference time domain algorithms to solve the wave equation in two and three dimensions -- Review of electromagnetic theory -- The Yee algorithm in one dimension -- The Yee algorithm in two and three dimensions -- Example applications of FDTD -- FDTD for dispersive materials -- Photonics problems -- Photonics design.
Subjects	Photonics--Mathematics. Electromagnetic waves--Mathematical models. Finite differences. Time-domain analysis.
Notes	Includes bibliographical references and index.

Dynamical Symmetry

LCCN	2010011912
Type of material	Book
Personal name	Wulfman, Carl.
Main title	Dynamical symmetry / Carl Wulfman.
Published/Created	Hackensack, N.J.: World Scientific, 2011.

Description	xx, 437 p.: ill.; 24 cm.
ISBN	9789814291361 (hardcover: alk. paper) 9814291366 (hardcover: alk. paper)
LC classification	QC174.17.S9 W85 2011
Contents	Physical symmetry and geometrical symmetry -- On symmetries associated with Hamiltonian dynamics -- One-parameter transformation groups -- Everywhere local invariance -- Lie transformation groups and Lie algebras -- Dynamical symmetry in Hamiltonian mechanics -- Symmetry in phase space -- Symmetries of classical Kepler motion -- Dynamical symmetry in Schrödinger quantum mechanics -- Spectrum generating groups -- Dynamical symmetry of regularized hydrogen-like atoms -- Approximate atomic and molecular calculations -- Rovibronic systems -- Symmetry of Maxwell's equations.
Subjects	Symmetry (Physics) Hamiltonian systems.
Notes	Includes bibliographical references and index.

Dynamics of Mechatronics Systems: Modeling, Simulation, Control, Optimization and experimental Investigations

LCCN	2016031840
Type of material	Book
Personal name	Awrejcewicz, J. (Jan), author.
Main title	Dynamics of mechatronics systems: modeling, simulation, control, optimization and experimental investigations / Jan Awrejcewicz, Donat Lewandowski, Pawel Olejnik, Lodz University of Technology, Poland.
Published/Produced	New Jersey: World Scientific, [2017]
Description	xiv, 342 pages; 25 cm
ISBN	9789813146549 (hardcover; alk. paper)

	9813146540 (hardcover; alk. paper)
LC classification	TJ163.12.A98 2017
Related names	Lewandowski, Donat, 1935- author. Olejnik, Pawel, 1975- author.
Summary	"This book describes the interplay of mechanics, electronics, electrotechnics, automation and biomechanics. It provides a broad overview of mechatronics systems ranging from modeling and dimensional analysis, and an overview of magnetic, electromagnetic and piezo-electric phenomena. It also includes the investigation of the pneumo-fluid-mechanical, as well as electrohydraulic servo systems, modeling of dynamics of an atom/particle embedded in the magnetic field, integrity aspects of the Maxwell's equations, the selected optimization problems of angular velocity control of a DC motor subjected to chaotic disturbances with and without stick-slip dynamics, and the analysis of a human chest adjacent to the elastic backrest aimed at controlling force to minimize relative compression of the chest employing the LQR. This book provides a theoretical background on the analysis of various kinds of mechatronics systems, along with their computational analysis, control, optimization as well as laboratory investigations"-- Provided by publisher.
Subjects	Mechatronics--Simulation methods. Mechatronics--Mathematical models. Machinery, Dynamics of.
Notes	Includes bibliographical references (pages 315-323) and index.

Earth's Magnetosphere: Formed by the Low-Latitude Boundary Layer

LCCN	2011293463
Type of material	Book
Personal name	Heikkila, Walter J.
Main title	Earth's magnetosphere: formed by the low-latitude boundary layer / Walter J. Heikkila.
Published/Created	Waltham, MA; Oxford, UK; Amsterdam, Netherlands: Elsevier, c2011.
Description	xviii, 477 p., [38] p. plates: ill. (some col.); 24 cm.
ISBN	9780444528643 (hbk.) 0444528644 (hbk.)
LC classification	QC809.M35 H45 2011
Summary	The author argues that, after four decades of debate about the interaction of solar wind with the magnetosphere, it is time to get back to basics. Starting with Newton's law, this book also examines Maxwell's equations and subsidiary equations such as continuity, constitutive relations and the Lorentz transformation, Helmholtz' theorem, and Poynting's theorem, among other methods for understanding this interaction. Includes chapters on prompt particle acceleration to high energies, plasma transfer event, and the low latitude boundary layer. More than 200 figures illustrate the text, which includes a color insert.
Contents	Prologue -- Historical introduction -- Approximate methods -- Helmholtz's theorem -- Poynting's energy conservation theorem -- Magnetopause -- High altitude cusps -- Low-latitude boundary layer -- Driving the plasma sheet -- Magnetospheric substorms -- Epilogue: What is new in this book.
Subjects	Magnetosphere. Magnetospheric boundary layer.

	Planets--Magnetospheres. Magnetosphäre. Earth (Planet)
Notes	Includes bibliographical references (p. [447]-471) and index.

Electricity and Magnetism: New Formulation by Introduction of Superconductivity

LCCN	2013951140
Type of material	Book
Personal name	Matsushita, Teruo, 1948- author.
Uniform title	Shin denjikigaku. English
Main title	Electricity and magnetism: new formulation by introduction of superconductivity / Teruo Matsushita.
Published/Produced	Tokyo: Springer, [2014]
Description	384 pages: illustrations; 23 cm.
ISBN	9784431545255 (pbk: acid-free paper) 4431545255 (pbk: acid-free paper)
LC classification	QC611.92.M3813 2014
Related names	Matsushita, Teruo, 1948- Shin denjikigaku. Translation of:
Summary	"The author introduces the concept that superconductivity can establish a perfect formalism of electricity and magnetism. The correspondence of electric materials that exhibit perfect electrostatic shielding (E=0) in the static condition and superconductors that show perfect diamagnetism (B=0) is given to help readers understand the relationship between electricity and magnetism. Another helpful aspect with the introduction of the superconductivity feature perfect diamagnetism is that the correspondence in the development of the expression of magnetic

energy and electric energy is clearly shown. Additionally, the basic mathematical operation and proofs are shown in an appendix, and there is full use of examples and exercises in each chapter with thorough answers." -- Back cover.

Contents Part I. Static electric phenomena: Electrostatic field; Conductors; Conductor systems in vacuum; Dielectric materials; Steady current -- Part II. Static magnetic phenomena: Current and magnetic flux density; Superconductors; Current systems; Magnetic materials -- Part III. Time-dependent electromagnetic phenomena: Electromagnetic induction; Displacement current and Maxwell's equations; Electromagnetic wave -- Appendices.

Subjects Superconductivity.
Electricity.
Magnetism.
Electromagnetism.
Electricity.
Electromagnetism.
Magnetism.
Superconductivity.

Notes Includes bibliographical references and index.
In English.

Additional formats Originally published in Japanese: Tōkyō: Koronasha, 2004, under title: Shin denjikigaku: denki jikigaku no atarashii taikei no kakuritsu. 9784339007640

Series Undergraduate lecture notes in physics; 2192-4791
Undergraduate lecture notes in physics.

Electricity and Magnetism for Mathematicians: A Guided Path from Maxwell's Equations to Yang-Mills

LCCN	2014035298
Type of material	Book
Personal name	Garrity, Thomas A., 1959- author.
Main title	Electricity and magnetism for mathematicians: a guided path from Maxwell's equations to Yang-Mills / Thomas A. Garrity, Williams College, Williamstown, Massachusetts; with illustrations by Nicholas Neumann-Chun.
Published/Produced	New York, NY: Cambridge University Press, 2015.
Description	xiv, 282 pages: illustrations; 24 cm
ISBN	9781107078208 (hardback) 1107078202 (hardback) 9781107435162 (paperback) 1107435161 (paperback)
LC classification	QC670.G376 2015
Related names	Neumann-Chun, Nicholas, illustrator.
Subjects	Electromagnetic theory--Mathematics--Textbooks. Science / Mathematical Physics.
Notes	Includes bibliographical references (pages 275-278) and index.

Electricity and Magnetism

LCCN	2012034622
Type of material	Book
Personal name	Purcell, Edward M.
Main title	Electricity and magnetism / Edward M. Purcell, David J. Morin, Harvard University, Massachusetts.
Edition	Third edition.
Published/Produced	Cambridge: Cambridge University Press, [2013]
Description	xxii, 839 pages: illustrations; 25 cm

Links	Cover image http://assets.cambridge.org/97811070/14022/cover/9781107014022.jpg
ISBN	9781107014022 (hardback)
LC classification	QC522.P85 2013
Summary	"For 50 years, Edward M. Purcell's classic textbook has introduced students to the wonders of electricity and magnetism. The third edition has been brought up to date and is now in SI units. It features hundreds of new examples, problems and figures and contains discussions of real-life applications. The textbook covers all the standard introductory topics, such as electrostatics, magnetism, circuits, electromagnetic waves and electric and magnetic fields in matter. Taking a non-traditional approach, magnetism is derived as a relativistic effect. Mathematical concepts are introduced in parallel with the physics topics at hand, making the motivations clear. Macroscopic phenomena are derived rigorously from microscopic phenomena. With worked examples, hundreds of illustrations and nearly 600 end-of-chapter problems and exercises, this textbook is ideal for electricity and magnetism courses. Solutions to the exercises are available for instructors at www.cambridge.org/9781107014022"-- Provided by publisher.
Contents	1. Electrostatics: charges and fields; 2. The electric potential; 3. Electric fields around conductors; 4. Electric currents; 5. The fields of moving charges; 6. The magnetic field; 7. Electromagnetic induction; 8. Alternating-current circuits; 9. Maxwell's equations and electromagnetic waves; 10. Electric fields in matter; 11. Magnetic fields in matter; Appendixes;

	References; Index.
Subjects	Electricity. Magnetism. Science / Physics.

Electricity and Magnetism

LCCN	2012392256
Type of material	Book
Personal name	Purcell, Edward M.
Main title	Electricity and magnetism / Edward M. Purcell.
Edition	2nd ed.
Published/Created	Cambridge; New York: Cambridge University Press, 2011.
Description	xix, 484 p.: ill.; 24 cm.
Links	Contributor biographical information http://www.loc.gov/catdir/enhancements/fy1209/2012392256-b.html Publisher description http://www.loc.gov/catdir/enhancements/fy1209/2012392256-d.html Table of contents only http://www.loc.gov/catdir/enhancements/fy1209/2012392256-t.html
ISBN	9781107013605
LC classification	QC522.P85 2011
Contents	Electrostatics: charges and fields -- The electric potential -- Electric fields around conductors -- Electric currents -- The fields of moving charges -- The magnetic field -- Electromagnetic induction -- Alternating-current circuits -- Maxwell's equations and electromagnetic waves -- Electric fields in matter -- Magnetic fields in matter.
Subjects	Electricity. Magnetism.
Notes	"Previously published by McGraw-Hill, Inc 1985. First published by Cambridge University Press

2011"--T.p. verso.
Includes index.

Electricity and Magnetism

LCCN	2014039158
Type of material	Book
Personal name	Nayfeh, Munir H. (Munir Hasan)
Main title	Electricity and magnetism / Munir H. Nayfeh and Morton K. Brussel, University of Illinois at Urbana-Champaign.
Edition	Dover edition.
Published/Produced	Mineola, New York: Dover Publications, Inc., 2015.
Description	xiv, 619 pages: illustrations; 24 cm
ISBN	9780486789712 (paperback) 0486789713 (paperback)
LC classification	QC522.N43 2015
Related names	Brussel, Morton K. Brussel, Morton K.
Summary	"This outstanding text for a two-semester course is geared toward physics undergraduates who have completed a basic first-year physics course. The coherent treatment offers several notable features, including 300 detailed examples at various levels of difficulty, a self-contained chapter on vector algebra, and a single chapter devoted to radiation that cites interrelationships between various analysis methods. Starting with chapters on vector analysis and electrostatics, the text covers electrostatic boundary value problems, formal and microscopic theories of dielectric electrostatics and of magnetism and matter, electrostatic energy, steady currents, and induction. Additional topics include magnetic energy, circuits with nonsteady

	currents, Maxwell's equations, radiation, electromagnetic boundary value problems, and the special theory of relativity. Exercises appear at the end of each chapter and answers to odd-numbered problems are included in one of several helpful appendixes"-- Provided by publisher.
Subjects	Electricity. Magnetism. Electromagnetic fields. Science / Electromagnetism. Technology & Engineering / Electrical.
Notes	Reprint of: New York: Wiley, ©1985. Includes bibliographical references and index.

Electromagnetic Computation Methods for Lightning Surge Protection Studies

LCCN	2015035649
Type of material	Book
Personal name	Baba, Yoshihiro, author.
Main title	Electromagnetic computation methods for lightning surge protection studies / Yoshihiro Baba, Vladimir A. Rakov.
Published/Produced	Singapore: IEEE, Wiley, 2016.
Description	xii, 315 pages; 26 cm
Links	Cover image http://catalogimages.wiley.com /images/db/jimages/9781118275634.jpg
ISBN	9781118275634 (hardback)
LC classification	TK3226.B2184 2016
Related names	Rakov, Vladimir A., 1955- author.
Summary	"Presents current research into electromagnetic computation theories with particular emphasis on Finite-Difference Time-Domain Method. This book is the first to consolidate current research and to examine the theories of electromagnetic

computation methods in relation to lightning surge protection. The authors introduce and compare existing electromagnetic computation methods such as the method of moments (MOM), the partial element equivalent circuit (PEEC), the finite element method (FEM), the transmission-line modeling (TLM) method, and the finite-difference time-domain (FDTD) method. The application of FDTD method to lightning protection studies is a topic that has matured through many practical applications in the past decade, and the authors explain the derivation of Maxwell's equations required by the FDTD, and modeling of various electrical components needed in computing lightning electromagnetic fields and surges with the FDTD method. The book describes the application of FDTD method to current and emerging problems of lightning surge protection of continuously more complex installations, particularly in critical infrastructures of energy and information, such as overhead power lines, air-insulated sub-stations, wind turbine generator towers and telecommunication towers. Both authors are internationally recognized experts in the area of lightning study and this is the first book to present current research in lightning surge protection Examines in detail why lightning surges occur and what can be done to protect against them Includes theories of electromagnetic computation methods and many examples of their application Accompanied by a sample printed program based on the finite-difference time-domain (FDTD) method written in C++ program "-- Provided by publisher.

Contents	Preface 1. Introduction 2. Lightning 3. The Finite-Difference Time-Domain Method for Solving Maxwell's Equations 4. Applications to Lightning Surge Protection Studies Appendix 3D-FDTD Program in C++ Index.
Subjects	Transients (Electricity)--Mathematical models. Lightning-arresters--Mathematical models. Lightning protection--Mathematical models. Electromagnetism--Mathematics. Time-domain analysis. Science / Electromagnetism.
Notes	Includes bibliographical references and index.
Additional formats	Online version: Baba, Yoshihiro, author. Electromagnetic computation methods for lightning surge protection studies Hoboken: John Wiley & Sons Inc., 2016 9781118275641 (DLC) 2015041277

Electromagnetic Fields and Waves

LCCN	2015304864
Type of material	Book
Personal name	Iskander, Magdy F.
Main title	Electromagnetic fields and waves / Magdy F. Iskander, University of Hawaii at Manoa.
Edition	Second edition.
Published/Produced	Long Grove, Illinois: Waveland Press, Inc. [2013] ©2013
Description	xxi, 906 pages: illustrations; 25 cm
ISBN	9781577667834 (hbk) 1577667832 (hbk)
LC classification	QC665.E4 I84 2013
Contents	Vector analysis and Maxwell's equations in integral form -- Maxwell's equations in differential form -- Maxwell's equations and plane wave propagation in

	materials -- Static electric and magnetic fields -- Normal incidence plane wave reflection and transmission at plane boundaries -- Oblique incidence plane wave reflection and transmission -- Transmission lines -- Waveguides -- Antennas.
Subjects	Electromagnetic fields. Electromagnetic waves.
Notes	Includes index.

Electromagnetics through the Finite Element Method: A Simplified Approach Using Maxwell's Equations

LCCN	2016020302
Type of material	Book
Personal name	Cardoso, José Roberto.
Main title	Electromagnetics through the finite element method: a simplified approach using Maxwell's equations / José Roberto Cardoso, Escola Politecnica da Universidade de Sao Paulo.
Published/Produced	Boca Raton: CRC Press, Taylor & Francis Group, [2017]
Description	xiii, 198 pages: illustrations (some color); 24 cm
ISBN	9781498783576 (hbk.: acid-free paper)
LC classification	QC760.C358 2017
Subjects	Electromagnetism--Mathematics. Maxwell equations--Numerical solutions. Finite element method.
Notes	Includes bibliographical references and index.

Engineered Materials and Metamaterials: Design and Fabrication

LCCN	2016013540
Type of material	Book
Personal name	Dudley, Richard A., 1979- author.
Main title	Engineered materials and metamaterials: design and fabrication / Richard A. Dudley Michael A.

	Fiddy.
Published/Produced	Bellingham, Washington: SPIE, [2017]
Description	xv, 203 pages: colored illustrations; 24 cm.
ISBN	9781510602151 (print; alk. paper)
	1510602151 (print; alk. paper)
	(pdf)
	(pdf)
	(epub)
	(epub)
	(Kindle/mobi)
	(Kindle/mobi)
LC classification	TK7871.15.M48 D83 2017
Related names	Fiddy, M. A., author.
Summary	"The field of metamaterials arose from a deeper understanding of how electromagnetic waves interact with materials and subwavelength-scaled scattering structures. This opened up the field of metamaterials or engineered materials through advances in understanding how material properties not found in nature could be designed along with advances in fabrication capabilities. Metamaterial advances span the electromagnetic spectrum, with examples being more common at lower (e.g., microwave) frequencies. The microwave or x-band regime has proven to be a good testbed for the first generation of metamaterials, but recently we have seen optical and IR metamaterials emerging as well. The exploitation of these more complex material-wave interactions, based on arrangements of subwavelength scale components, has generated a lot of global activity. We can, in principle, engineer material properties to greatly extend those currently available. This tutorial text presents both the usual and unusual electromagnetic properties of

materials, focusing especially man-made or engineered metamaterials. After a review of Maxwell's equations and material properties, the idea of resonant meta-atoms and composite media are introduced. The fabrication of metamaterials and the properties of negative index materials are explained. The difficulties associated with reducing the size of meta-atoms for use at optical frequencies are described, and the use of metamaterials for superresolution imaging is presented in some detail"-- Provided by publisher.

Contents	Material properties -- Meta-atoms -- Composite media and effective medium approximations -- Anisotropic microwave metamaterials -- Negative index -- Numerical simulations -- Making smaller structures: optical metamaterials -- Optical materials and fabrication challenges -- Superresolved imaging.
Subjects	Metamaterials. Optical materials. Electronics--Materials. Materials--Electric properties. Materials--Magnetic properties. High resolution imaging.
Notes	Includes bibliographical references and index.
Series	Tutorial texts in optical engineering; volume TT 106 Tutorial texts in optical engineering; v. TT 106.

Exercises for the FEYNMAN Lectures on Physics

LCCN	2014931781
Type of material	Book
Personal name	Feynman, Richard P. (Richard Phillips), 1918-1988.

Main title	Exercises for the Feynman lectures on physics / Richard Feynman, Robert Leighton, Matthew Sands [and others]; edited by Michael A. Gottlieb and Rudolf Pfeiffer.
Edition	New Millenium ed.
Published/Produced	New York: Basic Books, a member of the Perseus Group, [2014] ©2014
Description	1 volume (various pagings): illustrations; 28 cm
Links	Contributor biographical information http://www.loc.gov/catdir/enhancements/fy1511/2014931781-b.html Publisher description http://www.loc.gov/catdir/enhancements/fy1511/2014931781-d.html
ISBN	9780465060719 0465060714
LC classification	QC32.F39 2014
Related names	Leighton, Robert B. Sands, Matthew L. (Matthew Linzee)
Contents	Exercises for volume I. Atoms in motion -- Conservation of energy, statics -- Kepler's laws and gravitation -- Kinematics -- Newton's laws -- Conservation of momentum -- Vectors -- Non-relativistic two-body collisions in three dimensions -- Forces -- Potentials and fields -- Units and dimensions -- Relativistic kinematics and dynamics, mass and rest energy equivalence -- Relativistic energy and momentum -- Rotation in two dimensions, the center of mass -- Angular momentum, the moment of inertia -- Rotation in three dimensions -- The harmonic oscillator, linear differential equations -- Algebra -- Forced oscillations with damping -- Geometrical optics -- Electromagnetic radiation: interference --

Electromagnetic radiation: diffraction -- Electromagnetic radiation: refraction, dispersion, absorption -- Electromagnetic radiation: radiation damping, scattering -- Electromagnetic radiation: polarization -- Electromagnetic radiation: relativistic effects -- Quantum behavior: waves, particles, and photons -- Kinetic theory of gases -- Principles of statistical mechanics -- Applications of kinetic theory: equipartition -- Applications of kinetic theory: transport phenomena -- Thermodynamics -- Illustrations of thermodynamics -- The wave equation, sound -- Linear wave systems: beats, modes -- Fourier analysis of waves -- Exercises for volume II. Electromagnetism -- Differential calculus of vector fields -- Vector integral calculus -- Electrostatistics -- Applications of Gauss' Law -- The electric field in various circumstances -- The electric field in various circumstances (continued) -- Electrostatic energy -- Dielectrics -- Inside dielectrics -- Electrostatic analogs -- Magentostatics -- The magnetic field in various situations -- The vector potential -- The laws of induction -- Solutions of Maxwell's equations in free space -- Solutions of Maxwell's equations with currents and charges -- AC circuits -- Cavity resonators -- Waveguides -- Electrodynamics in relativistic notation -- Lorentz transformations of the fields -- Field energy and field momentum -- Electromagnetic mass -- The motion of charges in electric and magnetic fields -- Refractive index of dense materials -- Reflection from surfaces -- The magnetism of matter -- Paramagnetism and magnetic resonance -- Ferromagnetism -- Elasticity -- The flow of dry

	water -- The flow of wet water -- Exercises for volume III. Probability amplitudes -- Identical particles -- Spin one -- Spin one-half -- The dependence of amplitudes on time -- The Hamiltonian matrix -- The ammonia maser -- Other two-state systems -- More two-state systems -- The hyperfine splitting in hydrogen -- Propagation in a crystal lattice -- Semiconductors -- The independent particle approximation -- The dependence of amplitudes on position -- Angular momentum -- The hydrogen atom and the periodic table.
Subjects	Physics--Problems, exercises, etc. Physics.
Form/Genre	Problems, exercises, etc.

Fiber Optic Sensing and Imaging

LCCN	2013941502
Type of material	Book
Main title	Fiber optic sensing and imaging / Jin U. Kang, editor.
Published/Produced	New York, NY: Springer, [2013]
Description	vii, 171 pages: illustrations (some color); 24 cm
ISBN	9781461474814 1461474817 9781461474821 1461474825
LC classification	TA1815.F523 2013
Related names	Kang, Jin U.
Summary	This book is designed to highlight the basic principles of fiber optic imaging and sensing devices. The editor has organized the book to provide the reader with a solid foundation in fiber optic imaging and sensing devices. It begins with an introductory chapter that starts from Maxwell's

equations and ends with the derivation of the basic optical fiber characteristic equations and solutions (i.e. fiber modes). Chapter 2 reviews most common fiber optic interferometric devices and Chapter 3 discusses the basics of fiber optic imagers with emphasis on fiber optic confocal microscope. The fiber optic interferometric sensors are discussed in detail in chapter 4 and 5. Chapter 6 covers optical coherence tomography and goes into the details of signal processing and systems level approach of the real-time OCT implementation. Also useful forms of device characteristic equations are provided so that this book can be used as a reference for scientists and engineers in the optics and related fields.

Contents	Optical Fibers / Jin U. Kang -- Fiber Optic Interferometric Devices / Utkarsh Sharma, Xing Wei -- Fiber Optic Imagers / Do-Hyun Kim, Jin U. Kang -- Optical Fiber Gratings for Mechanical and Bio-sensing / Young-Geun Han -- Sagnac Loop Sensors / Li Qian -- Principles of Optical Coherence Tomography / Kang Zhang, Jin U. Kang.
Subjects	Optical fiber detectors. Fiber optics. Fiber optics. Optical fiber detectors. Faseroptischer Sensor. Faseroptik. Optische Abbildung.
Notes	Includes bibliographical references.

Finite Element Modeling Methods for Photonics

LCCN	2015430726
Type of material	Book
Personal name	Rahman, B. M. Azizur, author.
Main title	Finite element modeling methods for photonics / B. M. Azizur Rahman, Arti Agrawal.
Published/Produced	Boston: Artech House, [2013] ©2013
Description	xv, 247 pages: illustrations; 24 cm
ISBN	1608075311 9781608075317
LC classification	TA347.F5 R335 2013
Related names	Agrawal, Arti, author.
Contents	1. Introduction -- 1.1. Significance of Numerical Methods -- 1.2. Numerical Methods -- 1.3. Maxwell's Equations and Boundary Conditions -- 1.3.1. Maxwell's Equations -- 1.3.2. Boundary Conditions across Material Interfaces -- 1.3.3. Boundary Conditions: Natural and Forced -- 1.3.4. Boundary Conditions: Truncation of Domains -- 1.4. Basic Assumptions of Numerical Methods and Their Applicability -- 1.4.1. Time Harmonic and Time-Dependent Solutions -- 1.4.2. The Wave Equations -- 1.4.3. Scalar and Vector Nature of the Equations/Solutions -- 1.4.4. Modal Solutions -- 1.4.5. Beam Propagation Methods -- 1.5. Choosing a Modeling Method -- 1.6. Finite-Element-Based Methods -- References -- 2. The Finite-Element Method -- 2.1. Basic Concept of FEM: Essence of FEM-based Formulations -- 2.2. Setting up the FEM -- 2.2.1. The Variational Approach -- 2.2.2. The Galerkin Method -- 2.3. Scalar and Vector FEM Formulations --

2.3.1. The Scalar Formulation -- 2.3.2. The Vector Formulation -- 2.4. Implementation of FEM -- 2.4.1. Flowchart of Main Steps in FEM -- 2.4.2. Meshing and Shape Functions -- 2.4.3. Shape Functions -- 2.4.4. Examples of Meshing -- 2.5. Formation of Element and Global Matrices -- 2.5.1. Mass and Stiffness Matrix Evaluation for First-order Triangular Elements -- 2.5.2. Mass and Stiffness Matrix Evaluation for Second-order Triangular Elements -- 2.5.3. Assembly of Global Matrices: Bandwidth and Sparsity of Matrices -- 2.5.4. Penalty Function Method for Elimination of Spurious Modes -- 2.6. Solution of the System of Equations -- 2.7. Implementation of Boundary Conditions -- 2.7.1. Natural Boundary Condition and Symmetry: Electric and Magnetic Wall -- 2.7.2. Absorbing Boundary Condition and Perfectly Matched Layer (PML) Boundary Condition -- 2.7.3. Periodic Boundary Conditions (PBC) -- 2.8. Practical Illustrations of FEM Applied to Photonic Structures/devices -- 2.8.1. The Rectangular Waveguide: Si Nanowire -- 2.8.2. Waveguide with a Circular Cross Section: Photonic Crystal Fiber (PCF) -- 2.8.3. Plasmonic Waveguides -- 2.8.4. Photonic Crystal Waveguide and Periodic Boundary Conditions -- 2.9. FEM Analysis of Bent Waveguides -- 2.10. Perturbation Analysis for Loss/gain in Optical Waveguides -- 2.10.1. Perturbation Method with the Scalar FEM -- 2.10.2. Perturbation Method with the Vector FEM -- 2.11. Accuracy and Convergence in FEM -- 2.11.1. Discretisation and Interpolation Errors in FEM Analysis -- 2.11.2. Element Shape Quality and the Stiffness Matrix -- 2.11.3. Error Dependence on Element Size, Order and Arrangement -- 2.11.4. Adaptive Mesh Refinement -- 2.12. Computer Systems and FEM Implementation -- References -- 3. Finite-Element Beam Propagation Methods -- 3.1. Introduction -- 3.2. Setting up BPM Methods -- 3.3. Vector FE-BPM with PML Boundary Conditions -- 3.3.1. Semi-vector and Scalar FE-BPM -- 3.3.2. Wide-angle FE-BPM -- 3.3.3. Paraxial FE-BPM -- 3.3.4. Implementation of the BPM and Stability -- 3.3.5. Practical Illustrations of FE-BPM applied to Photonic Structures/devices -- 3.4. Junction Analysis with FEM: The LSBR Method -- 3.4.1. Analysis of High Index Contrast Bent Waveguide -- 3.5. Bi-directional BPM -- 3.6. Imaginary Axis/distance BPM -- 3.6.1. Analysis of 3D Leaky Waveguide by the Imaginary Axis BPM -- References -- 4. Finite-Element Time Domain Method -- 4.1. Time Domain Numerical Methods -- 4.2. Finite-Element Time Domain (FETD) BPM Method -- 4.2.1. Wide Band and Narrow Band Approximations -- 4.2.2. Implementation of the FETD BPM Method: Implicit and Explicit Schemes -- 4.3. Practical Illustrations of FETD BPM Applied to Photonic Structures/devices -- 4.3.1. Optical Grating -- 4.3.2.90° Sharp Bends -- References -- 5. Incorporating Physical Effects within the Finite-Element Method -- 5.1. Introduction -- 5.2. The Thermal Model -- 5.2.1. Thermal Modeling of a VCSEL -- 5.3. The Stress Model -- 5.3.1. Stress Analysis of a Polarization Maintaining Bow-tie Fiber -- 5.4. The

	Acoustic Model -- 5.4.1. Acousto-optic Analysis of a Silica Waveguide -- 5.4.2. SBS Analysis of a Silica Nanowire -- 5.5. The Electro-optic Model -- 5.5.1. Analysis of a Lithium Niobate (LN) Electro-optic Modulator -- 5.6. Nonlinear Photonic Devices -- 5.6.1. Analysis of a Strip-loaded Nonlinear Waveguide -- 5.6.2. Analysis of a Nonlinear Directional Coupler -- 5.6.3. Analysis of Second Harmonic Generation in an Optical Waveguide -- References -- 6. FE-based Methods: The Present and Future Directions -- 6.1. Introduction -- 6.2. Salient Features of FE-based Methods -- 6.3. Future Trends and Challenges for FE-based Methods -- Appendix A Scalar FEM with Perturbation -- TE Modes -- TM Modes -- Appendix B Vector FEM with Perturbation -- Appendix C Green's Theorem.
Subjects	Finite element method. Photonics. Finite element method. Photonics.
Notes	Includes bibliographical references and index.

Foundations of Antenna Engineering: A Unified Approach for Line-of-Sight and Multipath

LCCN	2015451245
Type of material	Book
Personal name	Kildal, Per-Simon, 1951- author.
Main title	Foundations of antenna engineering: a unified approach for line-of-sight and multipath / Per-Simon Kildal.
Published/Produced	Boston: Artech House, [2015] ©2015
Description	xxi, 455 pages: illustrations; 26 cm

ISBN	1608078671 9781608078677 160807868X 9781608078684
LC classification	TK7871.6.K48 2015
Summary	"This is the first textbook that contains a holistic treatment of traditional antennas mounted on masts (Line-of-Sight antenna systems) and small antennas used on modern wireless devices that are subject to signal variations (fading) due to multipath propagation. The focus is on characterization and describing classical antennas by modern complex vector theory, thereby linking together many disciplines, such as electromagnetic theory, classical antenna theory, wave propagation, and antenna system performance."--Back cover.
Contents	1.1. Antenna Types and Classes -- 1.2. Brief History of Antennas and Analysis Methods -- 1.3. Terminology, Quantities, Units, and Symbols -- 1.3.1. Radiation or Scattering -- 1.3.2. Reflection, Refraction, and Diffraction -- 1.3.3. Rays, Waves, Phase Fronts, and Phase Paths -- 1.3.4. SI Units for Fields and Sources and Decibels -- 1.3.5. Symbols -- 1.4. Vector Notation and Coordinate Transformations -- 1.4.1. Some Vector Formulas -- 1.4.2. Coordinate Transformations -- 1.4.3. Dyads -- 1.5. Overview on EM Analysis Methods by S. Maci -- References -- 2.1. Time-Harmonic Electromagnetic Fields -- 2.2. Plane Waves and Their Polarization -- 2.2.1. Linear Polarization -- 2.2.2. Circular Polarization -- 2.2.3. Axial Ratio and Cross-Polarization -- 2.2.4. Example: Amplitude and Phase Errors in Circular

Polarization Excitations -- 2.2.5. Polarizer for Generating Circular Polarization -- 2.2.6. Example: Mismatch in Polarizer -- 2.3. Radiation Fields -- 2.3.1. Field Regions -- 2.3.2. Radiation Fields of Receiving Antennas -- 2.3.3. Far-Field Function and Radiation Intensity -- 2.3.4. Phase Reference Point and Fraunhofer Approximation -- 2.3.5. Polarization of Radiation Fields -- 2.3.6. Copolar and Cross-Polar Radiation Patterns -- 2.3.7. Phase Center -- 2.3.8. Total Radiated Power -- 2.3.9. Directive Gain and Directivity -- 2.3.10. Beamwidth -- 2.3.11. Cross-Polarization -- 2.3.12. Beam Efficiency -- 2.3.13. E- and H-Plane Patterns -- 2.3.14. Fourier Expansion of the Radiation Field -- 2.3.15. Example: Phase Reference Point for Asymmetric Phase Pattern -- 2.3.16. Example: Calculation of Phase Center of a Symmetric Beam -- 2.4. Rotationally Symmetric Antennas (BOR) -- 2.4.1. BOR0 Antennas with Rotationally Symmetric Radiation Fields -- 2.4.2. BOR1 Antennas -- 2.4.3. Example: Directivity of BOR1 Antenna with Low Sidelobes -- 2.4.4. Example: Directivity of BOR1 Antenna with High Far-Out Sidelobes -- 2.4.5. Example: BOR1 Antenna with Different E- and H-Plane Patterns -- 2.4.6. Example: BOR1 Antenna with Different E- and H-Plane Phase Patterns -- 2.5. System Characteristics of the Antenna -- 2.5.1. Antenna Gain -- 2.5.2. Aperture Efficiency and Effective Area -- 2.5.3. Friis Transmission Equation and the Radar Equation -- 2.5.4. Antenna Noise Temperature and G/T -- 2.5.5. Bandwidth -- 2.5.6. Tolerances -- 2.5.7. Environmental Effects -

- 2.5.8. Example: Noise Temperature and G/T -- 2.6. Equivalent Circuits of Single-Port Antennas -- 2.6.1. Transmitting Antennas -- 2.6.2. Impedance Matching to Transmission Line -- 2.6.3. Receiving Antenna -- 2.6.4. Conjugate Impedance Matching -- 2.6.5. Impedance and Reflection Coefficient Transformations -- 2.7. Periodic Reflection Coefficients -- 2.8. Equivalent Circuits of Multiport Array Antennas -- 2.9. Further Reading -- 2.10. Complementary Comments by S. Maci -- 2.11. Exercises -- References -- 3.1. Multipath Without Line of Sight (LOS) -- 3.1.1. Rayleigh Fading and CDF -- 3.1.2. Angle of Arrival (AoA), XPD, and Polarization Imbalance -- 3.1.3. Rich Isotropic Multipath (RIMP) -- 3.2. Characterization of Single-Port Antennas in RIMP -- 3.2.1. Antenna Impedance, Port Impedance, and Reflection Coefficient -- 3.2.2. Mean Effective Gain (MEG) and Mean Effective Directivity (MED) -- 3.2.3. Total Radiation Efficiency and Transmission Formula -- 3.3. Characterization of Multiport Antennas in RIMP -- 3.3.1. Definition of Channel -- 3.3.2. Embedded Elements -- 3.3.3. Embedded Radiation Efficiency and Decoupling Efficiency -- 3.3.4. Correlation Between Ports -- 3.4. Characterization of Diversity Performance -- 3.4.1. Channel Estimation and Digital MRC Processing -- 3.4.2. Example: MRC Applied to 2-D Slot Antenna Case -- 3.4.3. Diversity Gains (Apparent, Effective, and Actual) -- 3.4.4. Theoretical Determination of Diversity Gain -- 3.5. Maximum Available Capacity from Shannon -- 3.5.1. Single-Port System -- 3.5.2. Parallel Channels in LOS --

3.5.3. Parallel Channels in Multipath -- 3.5.4. Normalization -- 3.5.5. Numerical Simulation of Channels in Multipath -- 3.6. Emulation of RIMP Using Reverberation Chamber -- 3.6.1. Mode Stirring (Mechanical, Platform, Polarization) -- 3.6.2. The S-Parameters of the Chamber and of the Antennas -- 3.6.3. Rayleigh Fading, Rician Fading, and AoA Distribution -- 3.6.4. Average Transmission Level (Hill's Formula) and Calibration -- 3.6.5. Frequency Stirring on Net Transfer Function -- 3.6.6. Number of Independent Samples and Accuracy -- 3.7. Measurements in Reverberation Chamber -- 3.7.1. Calibration and Characterizing Multiport Antennas -- 3.7.2. Radiated Power, Receiver Sensitivity, and Data Throughput -- 3.8. System Modeling Using Digital Threshold Receiver -- 3.8.1. The Digital Threshold Receiver -- 3.8.2. Modeling OFDM in LTE 4G System -- 3.8.3. Theoretical and Measured Results for i. i. d. Diversity Case -- 3.9. MIMO Multiplexing to Obtain Multiple Bitstreams -- 3.9.1. Diagonalizing the Channel Matrix -- 3.9.2. Measurements of Two Bitstreams in Reverberation Chamber -- 3.9.3. Quality of Throughput in Terms of MIMO Efficiency -- 3.10. Antennas for Use on Handsets -- 3.11. Exercises -- References -- 4.1. Maxwell's Equations -- 4.1.1. Differential Form -- 4.1.2. Standard Boundary Conditions -- 4.1.3. Impressed Current Sources on PECs -- 4.1.4. Soft and Hard Boundary Conditions -- 4.1.5. Auxiliary Vector Potentials -- 4.2. Vector Integral Forms of the E- and H-Fields --

4.2.1. General Expressions -- 4.2.2. Radiating Far-Field Expressions -- 4.2.3. Duality -- 4.2.4. Superposition -- 4.2.5. Replacement Between Electric and Magnetic Currents -- 4.2.6. Frequency Scaling -- 4.3. Construction of Solutions: Uniqueness and Equivalence -- 4.3.1. PEC Equivalent and Magnetic Currents -- 4.3.2. Free Space and Huygens Equivalents -- 4.3.3. Physical Equivalent -- 4.4. Incremental Current Sources -- 4.4.1. Incremental Electric Current (or Hertz Dipole) -- 4.4.2. Incremental Magnetic Current -- 4.4.3. Huygens Source -- 4.4.4. Summary -- 4.4.5. Example: Directivities of Incremental Sources -- 4.5. Reaction, Reciprocity, and Mutual Coupling -- 4.5.1. Reaction Integrals -- 4.5.2. Three Reciprocity Relations -- 4.5.3. Reciprocity Between Input and Output Ports of Antennas -- 4.5.4. Mutual Impedance, Mutual Admittance, and Coupling Coefficient -- 4.6. Imaging -- 4.7. Integral Equations, Method of Moments and Galerkin's Method -- 4.7.1. Simple Algorithm for the Near Field from the Line Current -- 4.7.2. Simple Algorithm for the Near Field from the Surface Current -- 4.8. Complementary Comments by S. Maci -- 4.9. Exercises -- References -- 5.1. Electric Monopole and Dipole -- 5.1.1. Approximate Current Distribution of a Monopole -- 5.1.2. Approximate Current Distribution of a Dipole -- 5.1.3. Far-Field Function of a Dipole -- 5.1.4. Directivity and Radiation Resistance of a Short Dipole -- 5.1.5. Equivalent Circuit and Maximum Effective Aperture of a Short Dipole -- 5.1.6. Directivity and Radiation Resistance of a

Half-Wave Dipole -- 5.1.7. Self-Impedance of an Electric Dipole -- 5.1.8. Impedance of Cylindrical and Flat Electric Dipoles -- 5.1.9. Dipole at an Arbitrary Location -- 5.1.10. Arbitrary Dipole Above Ground -- 5.1.11. Vertical Dipole Above Ground -- 5.1.12. Vertical Monopole -- 5.1.13. Horizontal Dipole Above Ground -- 5.2. Electric Loop Antenna as Vertical Magnetic Dipole -- 5.3. Helical Antennas -- 5.4. Slot Antennas -- 5.4.1. Field Distribution and Radiation Pattern -- 5.4.2. Slot Admittance When Excited by Voltage Source -- 5.4.3. Slot Excited by a Plane Wave -- 5.4.4. Reflection Coefficient of Open Waveguide -- 5.4.5. Slots in Waveguide Walls -- 5.5. Further Reading -- 5.6. Complementary Comments by S. Maci -- 5.7. Exercises -- References -- 6.1. Transmission Line Model for a Rectangular Patch -- 6.1.1. Radiation Pattern by a Two-Slot Model -- 6.1.2. Impedance by a Transmission Line Model -- 6.2. Self-Reaction Model for Patch Impedance -- 6.2.1. Expansion of Current Distribution and Method of Moment -- 6.2.2. Impedance of Line-Fed Patches -- 6.2.3. Impedance of Probe-Fed Patches -- 6.3. Spectral Domain Methods -- 6.3.1.3-D Field Problem -- 6.3.2. Harmonic 1-D Field Problem -- 6.3.3. Green's Function of Harmonic 1-D Field Problem -- 6.3.4. Numerical Implementation -- 6.4. Further Reading -- 6.5. Complementary Comments by S. Maci -- 6.6. Exercises -- References -- 7.1. Apertures in PECs -- 7.1.1. PECs of Arbitrary Shape -- 7.1.2. Infinite PEC Planes -- 7.2. Virtual Apertures in Free Space -- 7.2.1. Free Space and Huygens

Equivalents -- 7.2.2. Plane Apertures -- 7.3. Apertures in the xy-Plane -- 7.3.1. PEC Aperture and Its Incremental Element Factor -- 7.3.2. Free-Space Aperture and Its Incremental Element Factor -- 7.3.3. Power Integration over Aperture and Maximum Directivity -- 7.4. Rectangular Plane Aperture -- 7.4.1. E- and H-Plane Patterns -- 7.4.2. Directivity and Aperture Efficiency -- 7.4.3. Uniform Aperture Distribution -- 7.5. Circular Aperture with BOR1 Excitation -- 7.5.1. Aperture Field and Far-Field Function -- 7.5.2. Uniform Aperture Distribution -- 7.5.3. Gaussian Aperture Distribution -- 7.5.4. Tapered Aperture Distributions -- 7.6. Gaussian Beam -- 7.6.1. Gaussian Near Field -- 7.6.2. Phase Center of Gaussian Beam -- 7.6.3. Gaussian Far Field -- 7.6.4. Aperture Diffraction by Constant Phase Aperture -- 7.6.5. GO Radiation from Aperture with a Strongly Curved Wavefront -- 7.6.6. Alternative Expressions for Gaussian Beam Parameters -- 7.7. Complementary Comments -- 7.8. Exercises -- References -- 8.1. Calculation Methods -- 8.1.1. Cylindrical Waveguide Plane Aperture Approach -- 8.1.2. Radial Cylindrical Waveguide Approach -- 8.1.3. Conical and Apherical Sector Waveguide Approach -- 8.1.4. Flared Cylindrical Waveguide Approach -- 8.1.5. Mode-Matching Approach -- 8.1.6. Method of Moment Approach -- 8.2. E-Plane Sector Horn -- 8.2.1. Flared Cylindrical Waveguide Approach -- 8.2.2. Paraxial Approximation for a Plane Aperture Field -- 8.2.3. Radiation Patterns -- 8.3. H-Plane Sector Horn -- 8.3.1. Flared Cylindrical Waveguide

	Approach -- 8.3.2. Paraxial Approximation for a Plane Aperture Field -- 8.3.3. Radiation Patterns --
Subjects	Antennas (Electronics)
	Antennas (Electronics)
Notes	Includes bibliographical references and index.
Series	Artech House antennas and electromagnetics analysis library
	Artech House antennas and electromagnetics analysis library.

From Maxwell's Equations to Free and Guided Electromagnetic Waves: An Introduction for First-Year Undergraduates

LCCN	2014006532
Type of material	Book
Personal name	Quesada-Pérez, Manuel, author.
Main title	From Maxwell's equations to free and guided electromagnetic waves: an introduction for first-year undergraduates / Manuel Quesada-Pérez and José Alberto Maroto-Centeno.
Published/Produced	New York: Novinka, [2014]
	©2014
Description	ix,136 pages: illustrations; 23 cm.
ISBN	9781631174537 (soft cover)
	1631174533 (soft cover)
LC classification	QC661.Q47 2014
Related names	Maroto-Centeno, José Alberto.
Contents	Maxwell's equations in differential form -- Electromagnetic waves in free space -- Guided electromagnetic waves.
Subjects	Electromagnetic waves.
	Maxwell equations.
Notes	Includes bibliographical references (pages 135-136).
Series	Physics research and technology

Geometry and Light: The Science of Invisibility

LCCN	2010034728
Type of material	Book
Personal name	Leonhardt, Ulf, 1965-
Main title	Geometry and light: the science of invisibility / Ulf Leonhardt, Thomas Philbin.
Published/Created	Mineola, N.Y.: Dover Publications, c2010.
Description	vi, 278 p.: ill. (chiefly col.); 24 cm.
Links	Publisher description http://catdir.loc.gov/catdir/enhancements/fy1012/2010034728-d.html Table of contents only http://www.loc.gov/catdir/enhancements/fy1318/2010034728-t.html
ISBN	9780486476933 0486476936
LC classification	QA641.L435 2010
Related names	Philbin, Thomas, 1970-
Summary	"The science of invisibility combines two of physics' greatest concepts: Einstein's general relativity and Maxwell's principles of electromagnetism. Recent years have witnessed major breakthroughs in the area, and the authors of this volume - Ulf Leonhardt and Thomas Philbin of Scotland's University of St. Andrews - have been active in the transformation of invisibility from fiction into science. Their work on designing invisibility devices is based on modern metamaterials, inspired by Fermat's principle, analogies between mechanics and optics, and the geometry of curved space. Suitable for graduate students and advanced undergraduates of engineering, physics, or mathematics, and scientific researchers of all types, this is the first authoritative textbook on invisibility and the science behind it. The book is

	two books in one: it introduces the mathematical foundations - differential geometry - for physicists and engineers, and it shows how concepts from general relativity become practically useful in electrical and optical engineering, not only for invisibility but also for perfect imaging and other fascinating topics. More than one hundred full-color illustrations and exercises with solutions complement the text."--Publisher's description.
Contents	Prologue -- Fermat's principle -- Differential geometry -- Maxwell's equations -- Geometries and media.
Subjects	Geometry, Differential. Geometry--Foundations.
Notes	Includes bibliographical references and index.
Series	Dover books on physics Dover books on physics.

How to Solve Physics Problems

LCCN	2017448820
Type of material	Book
Personal name	Oman, Robert M., author.
Main title	How to solve physics problems / Daniel Milton Oman, PhD, Robert Milton Oman, PhD.
Edition	Second edition.
Published/Created	New York: McGraw-Hill Education, [2016]
Description	xiii, 416 pages: illustrations; 28 cm
ISBN	9780071849319 (pbk.) 0071849319 (pbk.)
LC classification	QC32.O53 2016
Related names	Oman, Daniel M., author.
Contents	Preface -- How to use this book -- How to excel in your physics course -- Mathematical background -- Vectors -- Kinematics in one dimension --

	Falling-body problems -- Projectile motion -- Forces (including centripetal force) -- Apparent weight -- Work and the definite integral -- Work-energy problems -- Center of mass and momentum -- Collision and impulse -- Rotational motion -- Rotational dynamics -- Equilibrium -- Gravity -- Simple harmonic motion -- Fluids -- Temperature and calorimetry -- Kinetics and the ideal gas laws -- First law of thermodynamics -- Second law of thermodynamics -- Mechanical waves -- Standing waves (strings and pipes) -- Sound -- Charge and Coulomb's law -- The electric field -- Gauss' law -- Electric potential -- Capacitance -- Current and resistance -- Resistors in DC circuits -- Kirchhoff's laws -- RC circuits -- Magnetic fields -- Magnetic forces -- Ampere's law -- Biot-Savart law -- Faraday's law -- Inductance -- RL circuits -- Oscillating LC circuits -- Series RLC circuits and phasors -- Maxwell's equations -- Electromagnetic waves -- Reflection, refraction, and polarization -- Mirrors and lenses -- Diffraction and interference -- Special relativity -- Quantum physics -- Atoms, molecules, and solids -- Nuclear physics -- Physical constants -- Conversions -- Periodic table of the elements.
Subjects	Physics--Problems, exercises, etc.
Notes	Includes index.

In Pursuit of the Unknown: 17 Equations that Changed the World

LCCN	2011944850
Type of material	Book
Personal name	Stewart, Ian, 1945-
Main title	In pursuit of the unknown: 17 equations that changed the world / Ian Stewart.

Published/Created	New York: Basic Books, 2012.
Description	x, 342 p.: ill.; 25 cm.
Links	Contributor biographical information http://www.loc.gov/catdir/enhancements/fy1211/2011944850-b.html Publisher description http://www.loc.gov/catdir/enhancements/fy1211/2011944850-d.html
ISBN	9780465029730 0465029736
LC classification	QA21.S834 2012
Portion of title	17 equations that changed the world
Contents	Why equations? -- The squaw on the hippopotamus: Pythagoras's theorem -- Shortening the proceedings: logarithms -- Ghosts of departed quantities: calculus -- The system of the world: Newton's law of gravity -- Portent of the ideal world: the scare root of minus one -- Much ado about knotting: Euler's formula for polyhedra -- Patterns of chance: normal distribution -- Good vibrations: wave equation -- Ripples and blips: Fourier transform -- The ascent of humanity: Navier-Stokes equation -- Waves in the ether: Maxwell's equations -- Law and disorder: second law of thermodynamics -- One thing is absolute: relativity -- Quantum weirdness: Schrödinger equation -- Codes, communications, and computers: information theory -- The imbalance of nature: chaos theory -- The Midas formula: Black-Scholes equation -- Where next?
Subjects	Mathematics--History. Equations--History. Physics--History.
Notes	Includes bibliographical references (p. [321]-330) and index.

Introduction to Electricity and Magnetism: Solutions to Problems

LCCN	2019007559
Type of material	Book
Personal name	Walecka, John Dirk, 1932- author.
Main title	Introduction to electricity and magnetism: solutions to problems / John Dirk Walecka (College of William and Mary, USA).
Published/Produced	Singapore; Hackensack, NJ: World Scientific Publishing Co. Pte. Ltd., [2019] ©2019
ISBN	9789811202636 (pbk.; alk. paper) 981120263X (pbk.; alk. paper)
LC classification	QC522.W355 2019
Summary	"The previously published textbook Introduction to Electricity and Magnetism provides a clear, calculus-based introduction to a subject that together with classical mechanics, quantum mechanics, and modern physics lies at the heart of today's physics curriculum. The lectures, although relatively concise, covers from Coulomb's law to Maxwell's equations and special relativity in a lucid and logical fashion. The book also contains an extensive set of accessible problems that enhances and extends the coverage, so the present book, as an aid to teaching and learning, provides the solutions to those problems"-- Provided by publisher.
Contents	Electricity -- Coulomb's law -- The electric field -- Gauss' law -- The electrostatic potential -- Electric energy -- Capacitors and dielectrics -- Currents and Ohm's law -- DC circuits -- Review of electricity -- Magnetism -- Vectors -- The magnetic force and field -- Ampere's law -- Electromagnetic induction -- Magnetic materials --

	Time-dependent circuits -- Review of magnetism -- Electromagnetism -- Maxwell's equations -- Waves -- Electromagnetic waves -- More electromagnetic waves -- The theory of special relativity -- Review of electromagnetism.
Subjects	Electricity--Problems, exercises, etc. Magnetism--Problems, exercises, etc.
Notes	Includes bibliographical references and index.

Introduction to Geometry and Relativity

LCCN	2013014127
Type of material	Book
Personal name	Mello, David C. (David Cabral), 1950-
Main title	Introduction to geometry and relativity / David C. Mello.
Published/Produced	New York: Nova Science Publishers, Inc., 2013.
Description	xiii, 364 pages: illustrations; 26 cm
ISBN	9781626185425 (hardcover)
LC classification	QA641.M455 2013
Contents	Introduction to Newtonian Spacetime -- The Structure of Newtonian Spacetime -- Maxwell's Equations -- Introduction to Special Relativity -- Introduction to Differential Geometry -- Tensor Calculus -- Introduction to Minkowski Space -- Maxwell's Equations in Minkowski Space -- Relativistic Hydrodynamics -- Foundations of General Relativity -- Predictions of General Relativity.
Subjects	Geometry, Differential. Relativity (Physics) Space and time.
Notes	Includes bibliographical references (pages 355-357) and index.

Introduction to the Physics of Waves

LCCN	2012025066
Type of material	Book
Personal name	Freegarde, Tim, 1965-
Main title	Introduction to the physics of waves / Tim Freegarde, University of Southampton.
Published/Produced	Cambridge: Cambridge University Press, 2013.
Description	xiv, 296 pages: illustrations; 26 cm
Links	Cover image http://assets.cambridge.org/97805211/97571/cover/9780521197571.jpg Contributor biographical information http://www.loc.gov/catdir/enhancements/fy1211/2012025066-b.html Publisher description http://www.loc.gov/catdir/enhancements/fy1211/2012025066-d.html Table of contents only http://www.loc.gov/catdir/enhancements/fy1211/2012025066-t.html
ISBN	9780521197571
LC classification	QC157.F74 2013
Summary	"Balancing concise mathematical analysis with the real-world examples and practical applications that inspire students, this textbook provides a clear and approachable introduction to the physics of waves. The author shows through a broad approach how wave phenomena can be observed in a variety of physical situations and explains how their characteristics are linked to specific physical rules, from Maxwell's equations to Newton's laws of motion. Building on the logic and simple physics behind each phenomenon, the book draws on everyday, practical applications of wave phenomena, ranging from electromagnetism to oceanography, helping to engage students and connect core theory with practice. Mathematical

derivations are kept brief and textual commentary provides a non-mathematical perspective. Optional sections provide more examples along with higher-level analyses and discussion. This textbook introduces the physics of wave phenomena in a refreshingly approachable way, making it ideal for first- and second-year undergraduate students in the physical sciences"-- Provided by publisher.

"Balancing concise mathematical analysis with the real-world examples and practical applications that inspire students, this textbook provides a clear and approachable introduction to the physics of waves. The author shows through a broad approach how wave phenomena can be observed in a variety of physical situations and explains how their characteristics are linked to specific physical rules, from Maxwell's equations to Newton's laws of motion. Building on the logic and simple physics behind each phenomenon, the book draws on everyday, practical applications of wave phenomena, ranging from electromagnetism to oceanography, helping to engage students and connect core theory with practice. Mathematical derivations are kept brief and textual commentary provides a non-mathematical perspective. Optional sections provide more examples along with higher-level analyses and discussion. This textbook introduces the physics of wave phenomena in a refreshingly approachable way, making it ideal for first and second-year undergraduate students in the physical sciences"-- Provided by publisher.

Contents Preface; 1. The essence of wave motion; 2. Wave equations and their solution; 3. Further wave

	equations; 4. Sinusoidal waveforms; 5. Complex wavefunctions; 6. Huygens wave propagation; 7. Geometrical optics; 8. Interference; 9. Fraunhofer diffraction; 10. Longitudinal waves; 11. Continuity conditions; 12. Boundary conditions; 13. Linearity and superpositions; 14. Fourier series and transforms; 15. Waves in three dimensions; 16. Operators for wave motions; 17. Uncertainty and quantum mechanics; 18. Waves from moving sources; 19. Radiation from moving charges; Appendix: vector mathematics; Index.
Subjects	Waves--Textbooks. Wave-motion, Theory of--Textbooks. Science / Physics.
Notes	Includes bibliographical references (pages 286-290) and index.

Lectures on Classical Electrodynamics

LCCN	2014019553
Type of material	Book
Personal name	Englert, Berthold-Georg, 1953- author.
Main title	Lectures on classical electrodynamics / Berthold-Georg Englert, National University of Singapore, Singapore.
Published/Produced	New Jersey: World Scientific, [2014]
Description	xvii, 239 pages: illustrations; 24 cm
ISBN	9789814596923 (hardcover: alk. paper) 9789814596930 (pbk.: alk. paper)
LC classification	QC631.E54 2014
Contents	Maxwell's equations -- Electromagnetic pulses -- Lorentz transformation -- 3+1 (dimensional notation) -- Action, reaction/interaction -- Retarded potentials -- Radiation fields -- Spectral properties of radiation -- Time-dependent spectral

	distribution -- Synchrotron radiation -- Scattering -- Diffraction.
Subjects	Electrodynamics.
Notes	Includes index.

Magnetics, Dielectrics, and Wave Propagation with MATLAB Codes

LCCN	2010010717
Type of material	Book
Personal name	Vittoria, C.
Main title	Magnetics, dielectrics, and wave propagation with MATLAB codes / Carmine Vittoria.
Published/Created	Boca Raton: CRC Press, c2011.
Description	xvii, 450 p., [4] p. of plates: ill. (some col.); 25 cm.
Links	Contributor biographical information http://www.loc.gov/catdir/enhancements/fy1112/2010010717-b.html
ISBN	9781439841990 1439841993
LC classification	QC760.4.M37 V58 2011
Contents	1. Review of Maxwell Equations and Units -- Maxwell Equations in MKS System of Units -- Major and Minor Magnetic Hysteresis Loops -- Tensor and Dyadic Quantities -- Maxwell Equations in Gaussian System of Units -- External, Surface, and Internal Electromagnetic Fields -- Problems -- Appendix 1.A. Conversion of Units -- References -- Solutions -- 2. Classical Principles of Magnetism -- Historical Background -- First Observation of Magnetic Resonance -- Definition of Magnetic Dipole Moment -- Magnetostatics of Magnetized Bodies -- Electrostatics of Electric Dipole Moment -- Relationship between B and H Fields -- General

Definition of Magnetic Moment -- Classical Motion of the Magnetic Moment -- Problems -- Appendix 2.A -- References -- Solutions -- 3. Introduction to Magnetism -- Energy Levels and Wave Functions of Atoms -- Spin Motion -- Intra-Exchange Interactions -- Heisenberg Representation of Exchange Coupling -- Multiplet States -- Hund Rules -- Spin-Orbit Interaction -- Lande gj-Factor -- Effects of Magnetic Field on a Free Atom -- Crystal Field Effects on Magnetic Ions -- Superexchange Coupling between Magnetic Ions -- Double Superexchange Coupling -- Ferromagnetism in Magnetic Metals -- Problems -- Appendix 3.A. Matrix Representation of Quantum Mechanics -- References -- Solutions -- 4. Free Magnetic Energy -- Thermodynamics of Noninteracting Spins: Paramagnets -- Ferromagnetic Interaction in Solids -- Ferrimagnetic Ordering -- Spinwave Energy -- Effects of Thermal Spinwave Excitations -- Free Magnetic Energy -- Single Ion Model for Magnetic Anisotropy -- Pair Model -- Demagnetizing Field Contribution to Free Energy -- Numerical Examples -- Cubic Magnetic Anisotropy Energy -- Uniaxial Magnetic Anisotropy Energy -- Problems -- References -- Solutions -- 5. Phenomenological Theory -- Smit and Beljers Formulation -- Examples of Ferromagnetic Resonance -- Simple Model for Hysteresis -- General Formulation -- Connection between Free Energy and Internal Fields -- Static Field Equations -- Dynamic Equations of Motion -- Microwave Permeability -- Normal Modes -- Magnetic Relaxation -- Free Energy of Multi-

Domains -- Problems -- References -- Solutions -- 6. Electrical Properties of Magneto-Dielectric Films -- Basic Difference between Electric and Magnetic Dipole Moments -- Electric Dipole Orientation in a Field -- Equation of Motion of Electrical Dipole Moment in a Solid -- Free Energy of Electrical Materials -- Magneto-Elastic Coupling -- Microwave Properties of Perfect Conductors -- Principles of Superconductivity: Type I -- Magnetic Susceptibility of Superconductors: Type I -- London's Penetration Depth -- Type-II Superconductors -- Microwave Surface Impedance -- Conduction through a Non-Superconducting Constriction -- Isotopic Spin Representation of Feynman Equations -- Problems -- Appendix 6.A -- References -- Solutions -- 7. Kramers-Kronig Equations -- Problems -- References -- Solutions -- 8. Electromagnetic Wave Propagation in Anisotropic Magneto-Dielectric Media -- Spinwave Dispersions for Semi-Infinite Medium -- Spinwave Dispersion at High k-Values -- The k = 0 Spinwave Limit -- Sphere -- Thin Films -- Needle -- Surface or Localized Spinwave Excitations -- Pure Electromagnetic Modes of Propagation: Semi-Infinite Medium -- Coupling of the Equation of Motion and Maxwell's Equations -- Normal Modes of Spinwave Excitations -- Magnetostatic Wave Excitations -- M Perpendicular to Film Plane -- H in the Film Plane -- Ferrite Bounded by Parallel Plates -- Problems -- Appendix 8.A -- Perpendicular Case -- In Plane Case -- References -- Solutions -- 9. Spin Surface Boundary Conditions -- A Quantitative Estimate of Magnetic

	Surface Energy -- Another Source of Surface Magnetic Energy -- Static Field Boundary Conditions -- Dynamic Field Boundary Conditions -- Applications of Boundary Conditions -- H T to the Film Plane -- H // to the Film Plane -- Electromagnetic Spin Boundary Conditions -- Problems -- Appendix 9.A -- Perpendicular Case -- In Plane Case -- References -- Solutions -- 10. Matrix Representation of Wave Propagation -- Matrix Representation of Wave Propagation in Single Layers -- (//) Case -- (T) Case -- The Incident Field -- Ferromagnetic Resonance in Composite Structures: No Exchange Coupling -- Ferromagnetic Resonance in Composite Structures: Exchange Coupling -- (T) Case -- Boundary Conditions -- (//) Case -- Boundary Conditions (// FMR) -- Problems -- Appendix 10.A -- Calculation of Transmission Line Parameters from [A] Matrix -- Microwave Response to Microwave Cavity Loaded with Magnetic Thin Film -- References -- Solutions.
Subjects	MATLAB. Magnetics--Mathematics. Dielectrics--Mathematics. Radio wave propagation--Mathematics. Electromagnetic waves--Mathematical models.
Notes	Includes bibliographical references and index.

Mathematical Methods for Physics: Using MATLAB and Maple

LCCN	2017960715
Type of material	Book
Personal name	Claycomb, J. R., author.
Main title	Mathematical methods for physics: using MATLAB and Maple / James R. Claycomb.

Published/Produced	Dulles, Virginia: Mercury Learning & Information, [2018]
Description	xx, 820 pages; 24 cm + 1 CD-ROM (4 3/4 in.)
ISBN	9781683920984 (hardcover) 1683920988 (hardcover)
LC classification	QC20.C615 2018
Contents	1.1. Algebra -- 1.1.1. Systems of Equations -- 1.1.2. Completing the Square -- 1.1.3. Common Denominator -- 1.1.4. Partial Fractions Decomposition -- 1.1.5. Inverse Functions -- 1.1.6. Exponential and Logarithmic Equations -- 1.1.7. Logarithms of Powers, Products and Ratios -- 1.1.8. Radioactive Decay -- 1.1.9. Transcendental Equations -- 1.1.10. Even and Odd Functions -- 1.1.11. Examples in Maple -- 1.2. Trigonometry -- 1.2.1. Polar Coordinates -- 1.2.2. Common Identities -- 1.2.3. Law of Cosines -- 1.2.4. Systems of Equations -- 1.2.5. Transcendental Equations -- 1.3. Complex Numbers -- 1.3.1. Complex Roots -- 1.3.2. Complex Arithmetic -- 1.3.3. Complex Conjugate -- 1.3.4. Euler's Formula -- 1.3.5. Complex Plane -- 1.3.6. Polar Form of Complex Numbers -- 1.3.7. Powers of Complex Numbers -- 1.3.8. Hyperbolic Functions -- 1.4. Elements of Calculus -- 1.4.1. Derivatives -- 1.4.2. Prime and Dot Notation -- 1.4.3. Chain Rule for Derivatives 1.4.4. Product Rule for Derivatives -- 1.4.5. Quotient Rule for Derivatives -- 1.4.6. Indefinite Integrals -- 1.4.7. Definite Integrals -- 1.4.8. Common Integrals and Derivatives -- 1.4.9. Derivatives of Trigonometric and Hyperbolic Functions -- 1.4.10. Euler's Formula -- 1.4.11. Integrals of Trigonometric and Hyperbolic Functions -- 1.4.12. Improper Integrals -- 1.4.13.

Integrals of Even and Odd Functions -- 1.5. MATLAB Examples -- 1.5.1. Functional Calculator -- 1.6. Exercises -- 2.1. Vectors and Scalars in Physics -- 2.1.1. Vector Addition and Unit Vectors -- 2.1.2. Scalar Product of Vectors -- 2.1.3. Vector Cross Product -- 2.1.4. Triple Vector Products -- 2.1.5. The Position Vector -- 2.1.6. Expressing Vectors in Different Coordinate Systems -- 2.2. Matrices in Physics -- 2.2.1. Matrix Dimension -- 2.2.2. Matrix Addition and Subtraction -- 2.2.3. Matrix Integration and Differentiation -- 2.2.4. Matrix Multiplication and Commutation -- 2.2.5. Direct Product2.2.6. Identity Matrix -- 2.2.7. Transpose of a Matrix -- 2.2.8. Symmetric and Antisymmetric Matrices -- 2.2.9. Diagonal Matrix -- 2.2.10. Tridiagonal Matrix -- 2.2.11. Orthogonal Matrices -- 2.2.12. Complex Conjugate of a Matrix -- 2.2.13. Matrix Adjoint (Hermitian Conjugate) -- 2.2.14. Unitary Matrix -- 2.2.15. Partitioned Matrix -- 2.2.16. Matrix Trace -- 2.2.17. Matrix Exponentiation -- 2.3. Matrix Determinant and Inverse -- 2.3.1. Matrix Inverse -- 2.3.2. Singular Matrices -- 2.3.3. Systems of Equations -- 2.4. Eigenvalues and Eigenvectors -- 2.4.1. Matrix Diagonalization -- 2.5. Rotation Matrices -- 2.5.1. Rotations in Two Dimensions -- 2.5.2. Rotations in Three Dimensions -- 2.5.3. Infinitesimal Rotations -- 2.6. MATLAB Examples -- 2.7. Exercises -- 3.1. Single-Variable Calculus -- 3.1.1. Critical Points -- 3.1.2. Integration with Substitution -- 3.1.3. Work-Energy Theorem -- 3.1.4. Integration by Parts -- 3.1.5. Integration with Partial Fractions3.1.6. Integration by Trig Substitution -- 3.1.7. Differentiating Across the Integral Sign -- 3.1.8.

Integrals of Logarithmic Functions -- 3.2. Multivariable Calculus -- 3.2.1. Partial Derivatives -- 3.2.2. Critical Points -- 3.2.3. Double Integrals -- 3.2.4. Triple Integrals -- 3.2.5. Orthogonal Coordinate Systems -- 3.2.6. Cartesian Coordinates -- 3.2.7. Cylindrical Coordinates -- 3.2.8. Spherical Coordinates -- 3.2.9. Line, Volume, and Surface Elements -- 3.3. Gaussian Integrals -- 3.3.1. Error Functions -- 3.4. Series and Approximations -- 3.4.1. Geometric Series -- 3.4.2. Taylor Series -- 3.4.3. Maclaurin Series -- 3.4.4. Index Labels -- 3.4.5. Convergence of Series -- 3.4.6. Ratio Test -- 3.4.7. Integral Test -- 3.4.8. Binomial Theorem -- 3.4.9. Binomial Approximations -- 3.5. Special Integrals -- 3.5.1. Integral Functions -- 3.5.2. Elliptic Integrals -- 3.5.3. Gamma Functions -- 3.5.4. Riemann Zeta Function -- 3.5.5. Writing Integrals in Dimensionless Form3.5.6. Black-Body Radiation -- 3.6. MATLAB Examples -- 3.7. Exercises -- 4.1. Vector and Scalar Fields -- 4.1.1. Scalar Fields -- 4.1.2. Vector Fields -- 4.1.3. Field Lines -- 4.2. Gradient of Scalar Fields -- 4.2.1. Gradient in Cartesian Coordinates -- 4.2.2. Unit Normal -- 4.2.3. Gradient in Curvilinear Coordinates -- 4.2.4. Cylindrical Coordinates -- 4.2.5. Spherical Coordinates -- 4.2.6. Scalar Field from the Gradient -- 4.3. Divergence of Vector Fields -- 4.3.1. Flux through a Surface -- 4.3.2. Divergence of a Vector Field -- 4.3.3. Gradient in Curvilinear Coordinates -- 4.3.4. Cylindrical Coordinates -- 4.3.5. Spherical Coordinates -- 4.4. Curl of Vector Fields -- 4.4.1. Line Integral -- 4.4.2. Curl of a Vector Field -- 4.4.3. Curl in Cartesian Coordinates -- 4.4.4. Curl in Curvilinear

Coordinates -- 4.4.5. Cylindrical Coordinates -- 4.4.6. Spherical Coordinates -- 4.4.7. Vector Potential -- 4.5. Laplacian of Scalar and Vector Fields4.5.1. Laplacian in Curvilinear Coordinates -- 4.5.2. Cylindrical Coordinates -- 4.5.3. Spherical Coordinates -- 4.5.4. The Vector Laplacian -- 4.6. Vector Identities -- 4.6.1. First Derivatives -- 4.6.2. First Derivatives of Products -- 4.6.3. Second Derivatives -- 4.6.4. Vector Laplacian -- 4.7. Integral Theorems -- 4.7.1. Gradient Theorem -- 4.7.2. Divergence Theorem -- 4.7.3. Cartesian Coordinates -- 4.7.4. Cylindrical Coordinates -- 4.7.5. Stokes's Curl Theorem -- 4.7.6. Navier-Stokes Equation -- 4.8. MATLAB Examples -- 4.9. Exercises -- 5.1. Classification of Differential Equations -- 5.1.1. Order -- 5.1.2. Degree -- 5.1.3. Solution by Direct Integration -- 5.1.4. Exact Differential Equations -- 5.1.5. Sturm-Liouville Form -- 5.2. First Order Differential Equations -- 5.2.1. Homogeneous Equations -- 5.2.2. Inhomogeneous Equations -- 5.3. Linear, Homogeneous with Constant Coefficients -- 5.3.1. Damped Harmonic Oscillator -- 5.3.2. Undamped Motion5.3.3. Overdamped Motion -- 5.3.4. Underdamped Motion -- 5.3.5. Critically Damped Oscillator -- 5.3.6. Higher Order Differential Equations -- 5.4. Linear Independence -- 5.4.1. Wronskian Determinant -- 5.5. Inhomogeneous with Constant Coefficients -- 5.6. Power Series Solutions to Differential Solutions -- 5.6.1. Standard Form -- 5.6.2. Airy's Differential Equation -- 5.6.3. Hermite's Differential Equation -- 5.6.4. Singular Points -- 5.6.5. Bessel's Differential Equation -- 5.6.6. Legendre's Differential Equation

-- 5.7. Systems of Differential Equations -- 5.7.1. Homogeneous Systems -- 5.7.2. Inhomogeneous Systems -- 5.7.3. Solution Vectors -- 5.7.4. Test for Linear Independence -- 5.7.5. General Solution of Homogeneous Systems -- 5.7.7. Charged Particle in Electric and Magnetic Fields -- 5.8. Phase Space -- 5.8.1. Phase Plots -- 5.8.2. Noncrossing Property -- 5.8.3. Autonomous Systems -- 5.8.4. Phase Space Volume -- 5.9. Nonlinear Differential Equations5.9.1. Predator-Prey System -- 5.9.2. Fixed Points -- 5.9.3. Linearization -- 5.9.4. Simple Pendulum -- 5.9.5. Numerical Solution -- 5.10. MATLAB Examples -- 5.11. Exercises -- 6.1. Dirac Delta Function -- 6.1.1. Representations of the Delta Function -- 6.1.2. Delta Function in Higher Dimensions -- 6.1.3. Delta Function in Spherical Coordinates -- 6.1.4. Poisson's Equation -- 6.1.5. Differential Form of Gauss's Law -- 6.1.6. Heaviside Step Function -- 6.2. Orthogonal Functions -- 6.2.1. Expansions in Orthogonal Functions -- 6.2.2. Completeness Relation -- 6.3. Legendre Polynomials -- 6.3.1. Associated Legendre Polynomials -- 6.3.2. Rodrigues' Formulas -- 6.3.3. Generating Functions -- 6.3.4. Orthogonality Relations -- 6.3.5. Spherical Harmonics -- 6.4. Laguerre Polynomials -- 6.4.1. Rodrigues' Formula -- 6.4.2. Generating Function -- 6.4.3. Orthogonality Relations -- 6.5. Hermite Polynomials -- 6.5.1. Rodrigues' Formula -- 6.5.2. Generating Function6.5.4. Orthogonality -- 6.6. Bessel Functions -- 6.6.1. Modified Bessel Functions -- 6.6.2. Generating Function -- 6.6.3. Spherical Bessel Functions -- 6.6.4. Rayleigh Formulas -- 6.6.5. Generating Functions -- 6.6.6.

Useful Relations -- 6.7. MATLAB Examples -- 6.8. Exercises -- 7.1. Fourier Series -- 7.1.1. Fourier Cosine Series -- 7.1.2. Fourier Sine Series -- 7.1.3. Fourier Exponential Series -- 7.2. Fourier Transforms -- 7.2.1. Power Spectrum -- 7.2.2. Spatial Transforms -- 7.3. Laplace Transforms -- 7.3.1. Properties of the Laplace Transform -- 7.3.2. Inverse Laplace Transform -- 7.3.3. Properties of Inverse Laplace Transforms -- 7.3.4. Table of Laplace Transforms -- 7.3.5. Solving Differential Equations -- 7.4. MATLAB Examples -- 7.5. Exercises -- 8.1. Types of Partial Differential Equations -- 8.1.1. First Order PDEs -- 8.1.2. Second Order PDEs -- 8.1.3. Laplace's Equation -- 8.1.4. Poisson's Equation -- 8.1.5. Diffusion Equation -- 8.1.6. Wave Equation8.1.7. Helmholtz Equation -- 8.1.8. Klein-Gordon Equation -- 8.2. The Heat Equation -- 8.2.1. Transient Heat Flow -- 8.2.2. Steady State Heat Flow -- 8.2.3. Laplace Transform Solution -- 8.3. Separation of Variables - - 8.3.1. The Heat Equation -- 8.3.2. Laplace's Equation in Cartesian Coordinates -- 8.3.3. Laplace's Equation in Cylindrical Coordinates -- 8.3.4. Wave Equation -- 8.3.5. Helmholtz Equation in Cylindrical Coordinates -- 8.3.6. Helmholtz Equation in Spherical Coordinates -- 8.4. MATLAB Examples -- 8.5. Exercises -- 9.1. Cauchy-Riemann Equations -- 9.1. Laplace's Equation -- 9.2. Integral Theorems -- 9.2.1. Cauchy's Integral Theorem -- 9.2.2. Cauchy's Integral Formula -- 9.2.3. Laurent Series Expansion -- 9.2.4. Types of Singularities -- 9.2.5. Residues -- 9.2.6. Residue Theorem -- 9.2.7. Improper Integrals -- 9.2.8. Fourier Transform Integrals -- 9.3.

Conformal Mapping -- 9.3.1. Poisson's Integral Formulas9.3.2. Schwarz-Christoffel Transformation -- 9.3.3. Conformal Mapping -- 9.3.4. Mappings on the Riemann Sphere -- 9.4. MATLAB Examples -- 9.5. Exercises -- 10.1. Velocity-Dependent Resistive Forces -- 10.1.1. Drag Force Proportional to the Velocity -- 10.1.2. Drag Force on a Falling Body -- 10.2. Variable Mass Dynamics -- 10.2.1. Rocket Motion -- 10.3. Lagrangian Dynamics -- 10.3.1. Calculus of Variations -- 10.3.2. Lagrange's Equations of Motion -- 10.3.3. Lagrange's Equations with Constraints -- 10.4. Hamiltonian Mechanics -- 10.4.1. Legendre Transformation -- 10.4.2. Hamilton's Equations of Motion -- 10.4.3. Poisson Brackets -- 10.5. Orbital and Periodic Motion -- 10.5.1. Kepler Problem -- 10.5.2. Periodic Motion -- 10.5.3. Small Oscillations -- 10.6. Chaotic Dynamics -- 10.6.1. Strange Attractors -- 10.6.2. Lorenz Model -- 10.6.3. Jerk Systems -- 10.6.4. Time Delay Coordinates -- 10.6.5. Lyapunov Exponents -- 10.6.6. Poincare Sections -- 10.7. Fractals10.7.1. Cantor Set -- 10.7.2. Koch Snowflake -- 10.7.3. Mandelbrot Set -- 10.7.4. Fractal Dimension -- 10.7.5. Chaotic Maps -- 10.8. MATLAB Examples -- 10.9. Exercises -- 11.1. Electrostatics in 1D -- 11.1.1. Integral and Differential Forms of Gauss's Law -- 11.1.2. Laplace's Equation in 1D -- 11.1.3. Poisson's Equation in 1D -- 11.2. Laplace's Equation in Cartesian Coordinates -- 11.2.1.3D Cartesian Coordinates -- 11.2.2. Method of Images -- 11.3. Laplace's Equation in Cylindrical Coordinates -- 11.3.1. Potentials with Planar Symmetry -- 11.3.2.

Potentials in 3D Cylindrical Coordinates -- 11.4. Laplace's Equation in Spherical Coordinates -- 11.4.1. Axially Symmetric Potentials -- 11.4.2.3D Spherical Coordinates -- 11.5. Multipole Expansion of Potential -- 11.5.1. Axially Symmetric Potentials -- 11.5.2. Off-Axis Trick -- 11.5.3. Asymmetric Potentials -- 11.6. Electricity and Magnetism -- 11.6.1. Comparison of Electrostatics and Magnetostatics11.6.2. Electrostatic Examples -- 11.6.3. Magnetostatic Examples -- 11.6.4. Static Electric and Magnetic Fields in Matter -- 11.6.5. Examples: Electrostatic Fields in Matter -- 11.6.6. Examples: Magnetic Fields in Matter -- 11.7. Scalar Electric and Magnetic Potentials -- 11.8. Time-Dependent Fields -- 11.8.1. The Ampere-Maxwell Equation -- 11.8.2. Maxwell's Equations -- 11.8.3. Self-Inductance -- 11.8.4. Mutual Inductance -- 11.8.5. Maxwell's Wave Equations -- 11.8.6. Maxwell's Equations in Matter -- 11.8.7. Time Harmonic Maxwell's Equations -- 11.8.8. Magnetic Monopoles -- 11.9. Radiation -- 11.9.1. Poynting Vector -- 11.9.2. Inhomogeneous Wave Equations -- 11.9.3. Gauge Transformation -- 11.9.4. Radiation Potential Formulation -- 11.9.5. The Hertz Dipole Antenna -- 11.10. MATLAB Examples -- 11.11. Exercises -- 12.1. Schrodinger Equation -- 12.1.1. Time-Dependent Schrodinger Equation -- 12.1.2. Time-Independent Schrodinger Equation12.1.3. Operators, Expectation Values and Uncertainty -- 12.1.4. Probability Current Density -- 12.2. Bound States I -- 12.2.1. Particle in a Box -- 12.2.2. Semi-Infinite Square Well -- 12.2.3. Square Well with a Step -- 12.3. Bound States II -- 12.3.1. Delta Function Potential -- 12.3.2. Quantum Bouncer --

12.3.3. Harmonic Oscillator -- 12.3.4. Operator Notation -- 12.3.5. Excited States of the Harmonic Oscillator -- 12.4. Schrodinger Equation in Higher Dimensions -- 12.4.1. Particle in a 3D Box -- 12.4.2. Schrodinger Equation in Spherical Coordinates -- 12.4.3. Radial Equation -- 12.4.4. Hydrogen Radial Wavefunctions -- 12.5. Approximation Methods -- 12.5.1. WKB Approximation -- 12.5.2. Time-Independent Perturbation Theory -- 12.5.3. Degenerate Perturbation Theory -- 12.5.4. Stark Effect -- 12.6. MATLAB Examples -- 12.7. Exercises -- 13.1. Microcanonical Ensemble -- 13.1.1. Number of Microstates and the Entropy -- 13.2. Canonical Ensemble13.2.1. Boltzmann Factor and Partition Function -- 13.2.2. Average Energy -- 13.2.3. Free Energy and Entropy -- 13.2.4. Specific Heat -- 13.2.5. Rigid Rotator -- 13.2.6. Harmonic Oscillator -- 13.2.7. Composite Systems -- 13.2.8. Stretching a Rubber Band -- 13.3. Continuous Energy Distributions -- 13.3.1. Partition Function and Average Energy -- 13.3.2. Particle in a Box -- 13.3.3. Maxwell-Boltzmann Distribution -- 13.3.4. Relativistic Gas -- 13.4. Grand Canonical Ensemble -- 13.4.1. Gibbs Factor -- 13.4.2. Average Energy and Particle Number -- 13.4.3. Single Species -- 13.4.4. Grand Potential -- 13.4.5. Comparison of Canonical and Grand Canonical Ensembles -- 13.4.6. Bose-Einstein Statistics -- 13.4.7. Black-Body Radiation -- 13.4.8. Debye Theory of Specific Heat -- 13.4.9. Fermi-Dirac Statistics -- 13.5. MATLAB Examples -- 13.6. Exercises -- 14.1. Kinematics -- 14.1.1. Postulates of Special Relativity -- 14.1.2. Time Dilatation -- 14.1.3.

Length Contraction14.1.4. Relativistic Doppler Effect -- 14.1.5. Galilean Transformation -- 14.1.6. Lorentz Transformations -- 14.1.7. Relativistic Addition of Velocities -- 14.1.8. Velocity Addition Approximation -- 14.1.9.4-Vector Notation -- 14.2. Energy and Momentum -- 14.2.1. Newton's Second Law -- 14.2.2. Mass Energy and Kinetic Energy -- 14.2.3. Low Velocity Approximation -- 14.2.4. Energy Momentum Relation -- 14.2.5. Completely Inelastic Collisions -- 14.2.6. Particle Decay -- 14.2.7. Energy Units -- 14.3. Electromagnetics in Relativity -- 14.3.1. Relativistic Transformation of Fields -- 14.3.2. Covariant Formulation of Maxwell's Equations -- 14.3.3. Homogeneous Maxwell Equations -- 14.3.4. Lorentz Force Equation -- 14.4. Relativistic Lagrangian Formulation -- 14.4.1. Lagrangian of a Free Particle -- 14.4.2. Relativistic 1D Harmonic Oscillator -- 14.4.3. Charged Particle in Electric and Magnetic Fields -- 14.5. MATLAB Examples -- 14.6. Exercises
15.1. The Equivalence Principle -- 15.1.1. Classical Approximation to Gravitational Redshift -- 15.1.2. Photon Emitted from a Spherical Star -- 15.1.3. Gravitational Time Dilation -- 15.1.4. Comparison of Time Dilation Factors -- 15.2. Tensor Calculus -- 15.2.1. Tensor Notation -- 15.2.2. Line Element and Spacetime Interval -- 15.2.3. Raising and Lowering Indices -- 15.2.4. Metric Tensor in Spherical Coordinates -- 15.2.5. Dot Product -- 15.2.6. Cross Product -- 15.2.7. Transformation Properties of Tensors -- 15.2.8. Quotient Rule for Tensors -- 15.2.9. Covariant Derivatives -- 15.3. Einstein's Equations -- 15.3.1. Geodesic Equations of Motion

	-- 15.3.2. Alternative Lagrangian -- 15.3.3. Riemann Curvature Tensor -- 15.3.4. Ricci Tensor -- 15.3.5. Ricci Scalar -- 15.3.6. Einstein Tensor -- 15.3.7. Einstein's Field Equations -- 15.3.8. Friedman Cosmology -- 15.3.9. Killing Vectors -- 15.4. MATLAB Examples -- 15.5. Exercises -- 16.1. Early Models -- 16.1.1. de Broglie waves 16.1.2. Klein-Gordon Equation -- 16.1.3. Probability Current Density -- 16.1.4. Lagrangian Formulation of the Klein-Gordon Equation -- 16.2. Dirac Equation -- 16.2.1. Derivation of a First Order Equation -- 16.2.2. Probability Current -- 16.2.3. Gamma Matrices -- 16.2.4. Positive and Negative Energies -- 16.2.5. Lagrangian Formulation of the Dirac Equation -- 16.3. Solutions to the Dirac Equation -- 16.3.1. Plane Wave Solutions -- 16.3.2. Nonplane Wave Solutions -- 16.3.3. Nonrelativistic Limit -- 16.3.4. Dirac Equation in an Electromagnetic Field -- 16.4. MATLAB Examples -- 16.5. Exercises.
Subjects	Mathematical physics. Mathematical physics.
Notes	Includes bibliographical references and index.

Mathematical Models and Numerical Simulation in Electromagnetism

LCCN	2013949325
Type of material	Book
Personal name	Bermúdez, Alfredo, author.
Main title	Mathematical models and numerical simulation in electromagnetism / Alfredo Bermúdez, Dolores Gómez, Pilar Salgado.
Published/Produced	Cham; New York: Springer, [2014]
Description	xvii, 432 pages: illustrations (chiefly color); 24 cm.

ISBN	9783319029481 3319029487
LC classification	QC760.B458 2014
Related names	Pedreira, Dolores Gomez, author. Salgado, Pilar, author.
Summary	The book represents a basic support for a master course in electromagnetism oriented to numerical simulation. The main goal of the book is that the reader knows the boundary-value problems of partial differential equations that should be solved in order to perform computer simulation of electromagnetic processes. Moreover it includes a part devoted to electric circuit theory based on ordinary differential equations. The book is mainly oriented to electric engineering applications, going from the general to the specific, namely, from the full Maxwell's equations to the particular cases of electrostatics, direct current, magnetostatics and eddy currents models. Apart from standard exercises related to analytical calculus, the book includes some others oriented to real-life applications solved with MaxFEM free simulation software.-- Source other than Library of Congress.
Contents	The harmonic oscillator -- The series RLC circuit -- Linear electronic circuits -- Maxwell's equation in free space -- Some solutions of Maxwell's equations in free space -- Maxwell's equations in material regions -- Electrostatics -- Direct current -- Magnetostatistics -- The eddy currents model -- An introduction to nonlinear magnetics. Hysteresis. -- Electrical circuits with MaxFEM -- Electrostatistics with MaxFEM -- Direct current with MaxFEM -- Magnetostatistics with MaxFEM

	-- Eddy currents with MaxFEM.
Subjects	Electromagnetism--Mathematical models.
	Electromagnetism--Mathematical models.
Notes	Includes bibliographical references (pages 421-425) and index.
Series	Unitext, 2038-3722; volume 74
	Unitext; v. 74.

Maxwell's Equations: Analysis and Numerics

LCCN	2019937571
Type of material	Book
Personal name	Langer, Ulrich.
Main title	Maxwell's equations: analysis and numerics / Ulrich Langer; [edited by] Ulrich Langer, Dirk Pauly, Sergey I. Repin.
Edition	1st edition.
Published/Produced	Boston, MA: De Gruyter, 2019.
ISBN	9783110542646 (print)
	9783110542691 (e-bk. (epub)
	9783110543612 (e-bk. (pdf)
Series	Radon Series on Computational and Applied Mathematics; 24

Maxwell's Equations of Electrodynamics: An Explanation

LCCN	2012040779
Type of material	Book
Personal name	Ball, David W. (David Warren), 1962-
Main title	Maxwell's equations of electrodynamics: an explanation / David W. Ball.
Published/Produced	Bellingham, Washington, USA: SPIE Press, [2012]
Description	x, 93 pages: illustrations; 22cm
ISBN	9780819494528 (pbk.)
LC classification	QC670.B27 2012
Subjects	Maxwell equations.

	Electromagnetic theory.
Notes	Includes bibliographical references (page 89) and index.

Maxwell's Equations

LCCN	2009019340
Type of material	Book
Personal name	Huray, Paul G., 1941-
Main title	Maxwell's equations / Paul G. Huray.
Published/Created	Hoboken, N.J.: Wiley: IEEE Press, c2010.
Description	xviii, 290 p.: ill. (some col.); 25 cm.
ISBN	9780470542767 (cloth)
	0470542764 (cloth)
LC classification	QC670.H87 2010
Subjects	Maxwell equations.
Notes	Includes bibliographical references.

Mixed Finite Element Methods and Applications

LCCN	2013940257
Type of material	Book
Personal name	Boffi, Daniele, author.
Main title	Mixed finite element methods and applications / Daniele Boffi, Franco Brezzi, Michel Fortin.
Published/Produced	Heidelberg; New York: Springer, [2013] ©2013
Description	xiv, 685 pages: illustrations; 25 cm.
ISBN	9783642365188 (alk. paper)
	3642365183 (alk. paper)
	9783642365195 (ebk.)
	3642365191 (ebk.)
LC classification	TA347.F5 B63 2013
Related names	Fortin, Michel, 1945- author.
	Brezzi, F. (Franco), 1945- author.
Summary	Non-standard finite element methods, in particular

mixed methods, are central to many applications. In this text the authors, Boffi, Brezzi and Fortin present a general framework, starting with a finite dimensional presentation, then moving on to formulation in Hilbert spaces and finally considering approximations, including stabilized methods and eigenvalue problems. This book also provides an introduction to standard finite element approximations, followed by the construction of elements for the approximation of mixed formulations in H(div) and H(curl). The general theory is applied to some classical examples: Dirichlet's problem, Stokes' problem, plate problems, elasticity and electromagnetism.

Contents Variational formulations and finite element methods -- Classical methods -- Model problems and elementary properties of some functional spaces -- Eigenvalue problems -- Duality methods -- Generalities -- Examples for symmetric problems -- Duality methods for non symmetric bilinear forms -- Mixed eigenvalue problems -- Domain decomposition methods, hybrid methods -- Modified variational formulations -- Augmented formulations -- Perturbed formulations -- Bibliographical remarks -- Function spaces and finite element approximations -- Properties of the spaces Hm(...), H(div;...), and H(curl:...) -- Basic properties -- Properties relative to a partition of... -- Properties relative to a change of variables -- De Rham diagram -- Finite element approximations of H1(...) and H2(...) -- Conforming methods -- Explicit basis functions on triangles and tetrahedra -- Nonconforming methods -- Quadrilateral finite elements on non affine meshes -- Quadrilateral approximation of scalar functions -- Non polynomial approximations -- Scaling arguments -- Simplicial approximations of H(div:...) and H(curl:...) -- Simplicial approximations of H(div:...) -- Simplicial approximation of H(curl:...) -- Approximations of H(div: K) on rectangles and cubes -- Raviart-Thomas elements on rectangles and cubes -- Other approximations of H(div: K) on rectangles -- Other approximations of H(div: K) on cubes -- Approximations of H(curl: K) on cubes -- Interpolation operator and error estimates -- Approximations of H(div: K) -- Approximation spaces for H(div:...) -- Approximations of H(curl:...) -- Approximation spaces for H(curl:...) -- Quadrilateral and hexahedral approximation of vector-valued functions in H(div:...) and H(curl:...) -- Discrete exact sequences -- Explicit basis functions for H(div: K) and H(curl: K) on triangles and tetrahedra -- Basis functions for H(div: K): the two-dimensional case -- Basis functions for H(div: K): the three-dimensional case -- Basis functions for H(curl: K): the two-dimensional case -- Basis functions for H(curl: K): the three-dimensional case -- Concluding remarks -- Algebraic aspects of saddle point problems -- Notation, and basic results in linear algebra -- Basic definitions -- Subspaces -- Orthogonal subspaces -- Orthogonal projections -- Basic results -- Restrictions of operators -- Existence and uniqueness of solutions: the solvability problem -- A preliminary discussion -- The necessary and sufficient condition -- Sufficient conditions -- Examples -- Composite matrices -- The solvability problem for perturbed matrices -- Preliminary results -- Main results -- Examples --

Stability -- Assumptions on the norms -- The inf-sup condition for the matrix b: an elementary discussion -- The inf-sup condition and the singular values -- The case of A elliptic on the whole space -- The case of A elliptic on the kernel of B -- The case of A satisfying an inf-sup on the kernel of B -- Additional results -- Some necessary conditions -- The case of B not surjective: modifikation of the problem -- Some special cases -- Composite matrices -- Stability of perturbed matrices -- The basic estimate -- The symmetric case for perturbed matrices -- Saddle point problems in hilbert spaces -- Reminders on hilbert spaces -- Scalar products, norms, completeness -- Closed subspaces and dense subspaces -- Orthogonality -- Continuous linear operators, dual spaces, polar spaces -- Bilinear forms and associated operators: transposed operators -- Dual spaces of linear subspaces -- Identification of a space with its dual space -- Restrictions of operators to closed subspaces -- Quotient spaces -- Existence and uniqueness of solutions -- Mixed formulations in Hilbert spaces -- Stability constants and inf-sup conditions -- The main result -- The case of lmB... Q' -- Examples -- Existence and uniqueness for perturbed problems -- Regular perturbations -- Singular perturbations -- Approximation of saddle point problems -- Basic results -- The basic assumptions -- The discrete operators -- Error estimates for finite dimensional approximations -- Discrete stability and error estimates -- Additional error estimates for the basic problem -- Variants of error estimates -- A simple example -- An important example: the pressure in the homogeneous stokes problem -- The case of

Ker Bth... (0) -- The case of Ker Bth... Ker Bt -- The case of Ker Bth... Ker Bt -- The case of... going to zero -- The inf-sup condition: criteria -- Some linguistic considerations -- General considerations -- The inf-sup condition and the B-compatible interpolation operator... -- Construction of... -- An alternative strategy: switching norms -- Extensions of error estimates -- Perturbed problems -- Penalty methods -- Singular perturbations -- Nonconforming methods -- Dual error estimates -- Numerical properties of the discrete problem -- The matrix form of the discrete problem -- And if the inf-sup condition does not hold? -- Solution methods -- Concluding remarks -- Complements: stabilisation methods, eigenvalue problems -- Augmented formulations -- An abstract framework for stabiiised methods -- Stabilising terms -- Stability conditions for augmented formulations -- Discretisations of augmented formulations -- Stabilising with the "element-wise equations" -- Other stabilisations -- General stability conditions -- Stability of discretised formulations -- Minimal stabilisations -- Another form of minimal stabilisation -- Enhanced strain methods -- Eigenvalue problems -- Some classical results. Eigenvalue problems in mixed form -- Special results for problems of Type (f, 0) and (0, g) -- Eigenvalue problems of the Type (o, g) -- Eigenvalue problems of the Form (0, g) -- Mixed methods for elliptic problems -- Non-standard methods for Dirichlet's problem -- Description of the problem -- Mixed finite element methods for Dirichlet's problem -- Eigenvalue problem for the mixed formulation -- Primal hybrid methods --

Primal macro-hybrid methods and domain decompositions -- Dual hybrid methods -- Numerical solutions -- Preliminaries -- Inter-element multipliers -- A brief analysis of the computational effort -- Error analysis for the multiplier -- Error estimates in other norms -- Application to an equation arising from semiconductor theory -- Using anisotropie meshes -- Relations with finite volume methods -- The one and two-dimensional cases -- The two-dimensional case -- The three-dimensional case -- Nonconforming methods: a trap to avoid -- Augmented formulations (Galerkin least squares methods) -- A posteriori error estimates -- Incompressible materials and flow problems -- Introduction -- The stokes problem as a mixed problem -- Mixed formulation -- Some examples of failure and empirical cures -- Continuous pressure: the... P1- P1 Element -- Discontinuous pressure: the P1-P0 Approximation -- Building a B-compatible operator: the simplest stable elements -- Building a B-compatible operator -- A stable case: the mini element -- Another stable approximation: the bi-dimensional P2-P0 element -- The nonconforming P1-P0 approximation -- Other techniques for checking the inf-sup condition -- Projection onto constants -- Verfürth's trick -- Space and domain decomposition techniques -- Macro-element technique -- Making use of the internal degrees of freedom -- Two-dimensional stable elements -- Continuous pressure elements -- Discontinuous pressure elements -- Quadrilateral elements, Qk-Pk-1 elements -- Three-dimensional stable elements -- Continuous pressure 3-d elements -- Discontinuous pressure 3-d elements -- Pk-Pk-1 schemes and generalised Hood-Taylor elements -- Discontinuous pressure Pk-Pk-1 elements -- Generalised Hood-Taylor elements -- Other developments for divergence-free stokes approximation and mass conservation -- Exactly divergence-free stokes elements, discontinuous Galerkin methods -- Stokes elements allowing for element-wise mass conservation -- Spurious pressure modes -- Living with spurious pressure modes: partial convergence -- The bilinear velocity-constant pressure Q1-P0 element -- Eigenvalue problems -- Nearly incompressible elasticity, reduced integration methods and relation with penalty methods -- Variational formulations and admissible discretisations -- Reduced integration methods -- Effects of inexact integration -- Other stabilisation procedures -- Augmented method for the stokes problem -- Defining an approximate inverse Sh-1 -- Minimal stabilisations for stokes -- Concluding remarks: choice of elements -- Choice of elements -- Complements on elasticity problems -- Introduction -- Continuous formulation of Stress methods -- Numerical approximations of Stress formulations -- Relaxed symmetry -- Tensors, tensorial notation and results on symmetry -- Continuous formulation of the relaxed symmetry approach -- Numerical approximation of relaxed-symmetry formulations -- Some families of methods with reduced symmetry -- Methods based on stokes elements -- Stabilisation by H(curl) bubbles -- Two examples -- Methods based on the properties of... -- Loosing the inclusion of kernel: stabiiised methods -- Concluding remarks --

Complements on plate problems -- A mixed fourth-order problem -- The... biharmonic problem -- Eigenvalues of the biharmonic problem -- Dual hybrid methods for plate bending problems -- Mixed methods for linear thin plates -- Moderately thick plates -- Generalities -- The mathematical formulation -- Mixed formulation of the Mindlin-Reissner model -- A decomposition principle and the stokes -- Connection -- Discretisation of the problem -- Continuous pressure approximations -- Discontinuous pressure elements -- Mixed finite elements for electromagnetic problems -- Useful results about the space H(curl:...), its boundary traces, and the de Rham complex -- The de Rham complex and the Helmholtz decomposition when... is simply connected -- The Friedrichs inequality -- Extension to more general topologies -- H(curl:...) In two space dimensions -- The time harmonic Maxwell system -- Maxwell's eigenvalue problem -- Analysis of the time harmonic Maxwell system -- Approximation of the time harmonic Maxwell equations -- Approximation of the Maxwell eigenvalue problem -- Analysis of the two-dimensional case -- Discrete compactness property -- Nodal finite elements -- Edge finite elements -- Enforcing the divergence-free condition by a penalty method -- Some remarks on exterior calculus -- Concluding remarks -- References -- Index.

Subjects Finite element method.
Finite element method--Data processing.
Computational Mathematics and Numerical Analysis.
Computational Science and Engineering.

	Theoretical and Applied Mechanics. Finite element method. Finite element method--Data processing. Finite-Elemente-Methode
Notes	Includes bibliographical references (pages 663-679) and index.
Additional formats	Printed edition: 9783642365188
Series	Springer series in computational mathematics, 0179-3632; 44 Springer series in computational mathematics; 44.

Modeling and Optimization of LCD Optical Performance

LCCN	2014001761
Type of material	Book
Personal name	Yakovlev, Dmitry A.
Main title	Modeling and optimization of LCD optical performance / Dmitry A. Yakovlev, Saratov State University, Russia, Vladimir G. Chigrinov, Hong Kong University of Science & Technology, Hong Kong, Hoi-Sing Kwok, Hong Kong University of Science & Technology, Hong Kong.
Published/Produced	Chichester, West Sussex, United Kingdom; Hoboken, NJ: John Wiley & Sons Inc., 2015.
Description	xvii, 554 pages: illustrations; 25 cm
Links	Cover image http://catalogimages.wiley.com/images/db/jimages/9780470689141.jpg
ISBN	9780470689141 (hardback)
LC classification	TK7872.L56 Y35 2015
Related names	Chigrinov, V. G. (Vladimir G.) Kwok, Hoi-Sing.
Summary	"The aim of this book is to present the theoretical foundations of modeling the optical characteristics of liquid crystal displays, critically reviewing modern modeling methods and examining areas of

applicability. The modern matrix formalisms of optics of anisotropic stratified media, most convenient for solving problems of numerical modeling and optimization of LCD, will be considered in detail. The benefits of combined use of the matrix methods will be shown, which generally provides the best compromise between physical adequacy and accuracy with computational efficiency and optimization facilities in the theoretical model. The book will include algorithms for solving common problems of LCD optics, and will give recommendations of how to build the basic theoretical model and choose mathematical tools to solve particular problems. Special attention will be paid to solving optimization and inverse problems of liquid crystal optics. Earlier books have covered the classic Jones Matrix method, but the authors will cover the newer, more universal and successful electrodynamic Jones matrix method; this has extremely high accuracy and is especially useful in oblique light incidence and because it acknowledges multiple reflection. This book will prove a useful tool for developers of new generations of liquid crystal displays, and for scientists dealing with optical investigation of liquid crystals. An appendix will be provided which includes a robust technique for calculating the equilibrium LC director field in 1D case"-- Provided by publisher.

Contents Preface 1 Polarization of monochromatic waves. Background of the Jones matrix methods. The Jones calculus 1.1 Homogeneous waves in isotropic media 1.1.1 Waves 1.1.2 Polarization. Jones vectors

1.1.3 Coordinate transformation rules for Jones vectors. Orthogonal polarizations. Decomposition of a wave into two orthogonally polarized waves 1.2 Interface optics for isotropic media 1.2.1 Fresnel's formulas. Snell's law 1.2.2 Reflection and transmission Jones matrices for a plane interface between isotropic media 1.3 Wave propagation in anisotropic media 1.3.1 Wave equations 1.3.2 Waves in a uniaxial layer 1.3.3 A simple birefringent layer. Principal axes of a simple birefringent layer 1.3.4 Transmission Jones matrices of a simple birefringent layer at normal incidence 1.3.5 Linear retarders 1.3.6 Jones matrices of absorbing polarizers 1.4 Jones calculus 1.4.1 Basic principles of the Jones calculus 1.4.2 Three useful theorems for transmissive systems 1.4.3 Reciprocity relations. Jones's reversibility theorem 1.4.4 Theorem of polarization reversibility for systems without diattenuation 1.4.5 Particular variants of application of the Jones calculus. Cartesian Jones vectors for wave fields in anisotropic media 2 The Jones calculus: Solutions for ideal twisted structures and their applications in LCD optics 2.1 Jones matrix and eigenmodes of a liquid crystal layer with an ideal twisted structure 2.2 LCD optics and Gooch-Tarry formulas 2.3 Interactive simulation 2.4 Parameter space 3 Optical equivalence theorem 3.1 General optical equivalence theorem 3.2 Optical equivalence for a twisted nematic liquid crystal cell 3.3 Polarization conserving modes 3.4 Application to nematic bistable LCDs 3.5 Application to reflective displays 3.6 Measurement of characteristic parameters of an LC cell 4 Electrooptical modes. Practical examples of LCD modeling and optimization 4.1 Optimization of LCD performance in various electrooptical modes 4.1.1 Electrically Controlled Birefringence 4.1.2 Twist effect 4.1.3 Supertwist effect 4.1.4 Optimization of optical performance of reflective LCDs 4.2 Transflective LCDs 4.2.1 Dual Mode Single Cell Gap approach 4.2.2 Single Mode Single Cell Gap approach 4.3 Total Internal Reflection Mode 4.4 Ferroelectric Liquid Crystal Displays 4.4.1 Basic physical properties 4.4.2 Electrooptic effects in FLC cells 4.5 Birefringent Color Generation in Dichromatic Reflective FLCDs 5 Necessary mathematics. Radiometric terms. Conventions. Various Stokes and Jones vectors 5.1 Some definitions and relations from matrix algebra 5.1.1 General definitions 5.1.2 Some important properties of matrix products 5.1.3 Unitary matrices. Unimodular unitary 2´ 2 matrices. STU matrices 5.1.4 Norms of vectors and matrices 5.1.5 Kronecker product of matrices 5.1.6 Approximations 5.2 Some radiometric quantities. Conventions 5.3 Stokes vectors of plane waves and collimated beams propagating in isotropic nonabsorbing media 5.4 Jones vectors 5.4.1 Fitted-to-electric-field Jones vectors and fitted-to-transverse-component-of-electric-field Jones vectors 5.4.2 Fitted-to-irradiance Jones vectors 5.4.3 Conventional Jones vectors 6 Simple models and representations for solving the optimization and inverse optical problems. Real optics of LC cells and useful approximations 6.1 Polarization transfer factor of an optical system 6.2 Optics of LC cells in terms of the polarization transport coefficients 6.2.1 Polarization-dependent losses, and depolarization, unpolarized transmittance 6.2.2 Rotations 6.2.3 Symmetry of sample 6.3 Retroreflection geometry 6.4 Applications of the polarization transport coefficients in the optimization of LC devices 6.5 Evaluation of the ultimate characteristics of an LCD that can be attained by fitting the compensation system. Modulation efficiency of LC layers 7 Some physical models and mathematical algorithms used in modeling the optical performance of LCDs 7.1 Physical models of the light - layered system interaction used in modeling the optical behavior of LC devices. Plane-wave approximations. Transfer channel approach 7.2 Transfer matrix technique and adding technique 7.2.1 Transfer matrix technique 7.2.2 Adding technique 7.3 Optical models of some elements of LCDs 8 Modeling methods based on the rigorous theory of the interaction of a plane monochromatic wave with an ideal stratified medium. Eigenwave (EW) methods. EW Jones matrix method 8.1 General properties of electromagnetic field induced by a plane monochromatic wave within a linear stratified media 8.1.1 Maxwell's equations and constitutive relations 8.1.2 Plane waves 8.1.3 Field geometry 8.2 Transmission and reflection operators of fragments (TR-units) of stratified medium and their calculation 8.2.1 EW Jones vector. EW Jones matrices. Transmission and reflection operators 8.2.2 Calculation of overall transmission and overall reflection operators for layered systems by using transfer matrices 8.3 Berreman's method 8.3.1 Transfer matrices 8.3.2 Transfer matrix of a homogeneous layer 8.3.3 Transfer matrix of a smoothly inhomogeneous layer. Staircase approximation 8.3.4 Coordinate systems 8.4 Simplifications, useful relations and advanced techniques 8.4.1 Orthogonality relations and other useful relations for eigenwave bases 8.4.2 Simple general formulas for transmission operators for interfaces 8.4.3 Calculation of transmission and reflection operators of layered systems by using the adding technique 8.5 Transmissivities and reflectivities 8.6 Mathematical properties of transfer matrices and 2´ 2 transmission and reflection matrices of lossless media and reciprocal media 8.6.1 Properties of matrix operators for nonabsorbing regions 8.6.2 Properties of matrix operators for reciprocal regions 8.7 Calculation of EW 4´4 transfer matrices for LC layers 8.8 Transformation of the elements of EW Jones vectors and EW Jones matrices under changes of eigenwave bases 8.8.1 Coordinates of the EW Jones vector of a wave field in different eigenwave bases 8.8.2 EW Jones operators in different eigenwave bases 9 Specification of eigenwave bases in isotropic, uniaxial, and biaxial media 9.1 General aspects of the eigenwave basis specification. Implementation topics 9.2 Isotropic media 9.3 Uniaxial media 9.4 Biaxial media 10 Efficient methods for calculating the optical characteristics of layered systems for quasimonochromatic incident light. Main routines of LMOPTICS library 10.1 EW Stokes vectors and EW Mueller matrices 10.2 Calculation of the EW Mueller matrices of the overall transmission and reflection of a system consisting of "thin" and "thick" layers 10.3 Main routines of LMOPTICS 10.3.1 Routines for computing 4´4 transfer matrices and EW Jones matrices 10.3.2 Routines for computing EW Mueller matrices 10.3.3 Other useful routines 11 Calculation of the transmission characteristics of inhomogeneous liquid crystal layers by using the classical Jones calculus and the EW Jones matrix method 11.1 Application of Jones matrix methods to modeling the optics of inhomogeneous LC layers 11.1.1 Calculation of a transmission Jones matrix of an LC layer by using the classical Jones calculus 11.1.2 Extended Jones matrix methods 11.2 NBR approximation. Basic differential equations 11.3 NBR approximation. Numerical methods 11.3.1 Approximating multilayer method 11.3.2 Discretization method 11.3.3 Power series method 11.4 NBR approximation. Analytical solutions 11.4.1 Twisted strictures 11.4.2 Non-twisted structures 11.4.3 NBR approximation and the geometrical optics approximation. Adiabatic and quasiadiabatic approximations. 11.5 Estimation of the maximum error of calculated values of the LCD panel transmittance caused by errors in used values of the transmission matrix of the LC layer 12 Some approximate representations in EW Jones matrix method and their application for solving optimization and inverse problems for LCDs 12.1 Theory of STUM approximation 12.2 Exact and approximate expressions for transmission operators of interfaces at normal incidence 12.3 Polarization Jones matrix of an inhomogeneous nonabsorbing anisotropic layer with negligible bulk reflection at normal incidence. Simple representations of the polarization matrices of LC layers at normal incidence 12.4 Immersion model of the

	polarization-converting system of an LCD 12.5 Determining the configurational and optical parameters of LC layers with a twisted structure: Spectral fitting method 12.5.1 How to bring together the experiment and unitary approximation 12.5.2 Parameterization and solving the inverse problem 12.5.3 Appendix to section 12.5 12.6 Development of compensation systems for enhancement of viewing angle characteristics of LCDs 13 A few words about modeling of fine-structure LCDs and the direct ray approximation 13.1 Virtual microscope 13.2 Directional illumination and diffuse illumination Appendix A A.1 Introductory remarks A.2 Fast LCD A.3 Color LCD A.4 Transflective LCD A.5 Switchable viewing angle LCD A.6 Optimal E-paper configurations A.7 Color filter optimization Appendix B B.1 Conservation law for energy flux B.2 Lorentz's lemma B.3 Non-exponential waves B.4 To the power series method (section 11.3.3) B.5 One of ways to obtain the explicit expressions for transmission Jones matrices of an ideal twisted LC layer.
Subjects	Liquid crystal displays. Technology & Engineering / Electronics / General.
Notes	Includes bibliographical references and index.
Additional formats	Online version: Yakovlev, Dmitry A. Modeling and optimization of LCD optical performance Chichester, West Sussex, United Kingdom; Hoboken, NJ: John Wiley & Sons Inc., 2014 9781118706718 (DLC) 2014011500

Myogenesis in Development and Disease

LCCN	2017471171
Type of material	Book
Main title	Myogenesis in development and disease / edited by David Sassoon.
Edition	First edition.
Published/Produced	Cambridge, MA; Academic Press, an imprint of Elsevier, 2018.
Description	xii, 322 pages: illustrations (some color); 24 cm
ISBN	9780128092156 (hbk.) 0128092157 (hbk.)
LC classification	QL979.M965 2018
Related names	Sassoon, D. A., editor.
Contents	"What Did Maxwell's Equations Really Have to Do With Edison's Invention?": Addressing the Complexity of Developing Clinical Interventions for Skeletal Muscle Disease / Jonathan Dando -- The Muscle Stem Cell Niche in Health and Disease / Omid Mashinchian, Addolorata Pisconti, Emmeran Le Moal and C. Florian Bentzinger -- Translational Control of the Myogenic Program in Developing, Regenerating, and Diseased Skeletal Muscle / Ryo Fujita and Colin Crist -- The Composition, Development, and Regeneration of Neuromuscular Junctions / Wenxuan Liu and Joe V. Chakkalakal -- Cellular Biomechanics in Skeletal Muscle Regeneration / Edward W. Li, Olivia C. McKee-Muir and Penney M. Gilbert -- Satellite Cell Self-Renewal / Lorenzo Giordani, Alice Parisi and Fabien Le Grand -- "Known Unknowns": Current Questions in Muscle Satellite Cell Biology / D.D.W. Cornelison -- Epigenetic Regulation of Adult Myogenesis / Daniel C.L. Robinson and Francis J. Dilworth -- Dysregulated

	Myogenesis in Rhabdomyosarcoma / Peter Y. Yu and Denis C. Guttridge -- Muscle Stem Cells and Aging / Ara B. Hwang and Andrew S. Brack.
Subjects	Myogenesis. Muscle Development. Myogenesis.
Notes	Includes bibliographical references.
Series	Current topics in developmental biology; volume 126 0070-2153 Current topics in developmental biology; v. 126. 0070-2153

Numerical Electromagnetics: The FDTD Method

LCCN	2010048442
Type of material	Book
Personal name	Inan, Umran S.
Main title	Numerical electromagnetics: the FDTD method / Umran S. Inan, Robert A. Marshall.
Published/Created	Cambridge; New York: Cambridge University Press, 2011.
Description	xiv, 390 p.: ill.; 26 cm.
Links	Cover image http://assets.cambridge.org/97805211/90695/cover/9780521190695.jpg
ISBN	9780521190695 (hardback) 052119069X (hardback)
LC classification	QC760.I589 2011
Related names	Marshall, Robert A.
Summary	"Beginning with the development of finite difference equations, and leading to the complete FDTD algorithm, this is a coherent introduction to the FDTD method (the method of choice for modeling Maxwell's equations). It provides students and professional engineers with everything they need to know to begin writing

FDTD simulations from scratch and to develop a thorough understanding of the inner workings of commercial FDTD software. Stability, numerical dispersion, sources and boundary conditions are all discussed in detail, as are dispersive and anisotropic materials. A comparative introduction of the finite volume and finite element methods is also provided. All concepts are introduced from first principles, so no prior modeling experience is required, and they are made easier to understand through numerous illustrative examples and the inclusion of both intuitive explanations and mathematical derivations"-- Provided by publisher.

Contents 1. Introduction; 2. Review of electromagnetic theory; 3. Partial differential equations and physical systems; 4. The FDTD grid and the Yee algorithm; 5. Numerical stability of finite difference methods; 6. Numerical dispersion and dissipation; 7. Introduction of sources; 8. Absorbing boundary conditions; 9. The perfectly matched layer; 10. FDTD modeling in dispersive media; 11. FDTD modeling in anistropic media; 12. Some advanced topics; 13. Unconditionally stable implicit FDTD methods; 14. Finite-difference frequency domain; 15. Finite volume and finite element methods.

Subjects Electromagnetism--Computer simulation.
Finite differences.
Time-domain analysis.

Notes Includes bibliographical references and index.

Optical Physics for Nanolithography

LCCN	2018930737
Type of material	Book
Personal name	Yen, Anthony, author.
Main title	Optical physics for nanolithography / Anthony Yen, Shinn-Sheng Yu.
Published/Produced	Bellingham, Washington: SPIE Press, [2018]
Description	336 pages: color illustrations; 28 cm.
ISBN	9781510617377 softcover 151061737X softcover pdf
LC classification	QC355.3.Y46 2018
Related names	Yu, Shinn-Sheng, author. Society of Photo-optical Instrumentation Engineers, publisher.
Summary	This book provides an in-depth, self-contained introduction of partially coherent imaging theory for researchers and engineers working on optical lithography for semiconductor manufacturing, including those in the EDA industry. It is mathematically complete: the opening chapters discuss the essential principles, and all derivations are presented with their intermediate steps. For increased accessibility, simplified and consistent notations are used throughout the text. Full-color pages illustrate the connections between figures and equations.-- Source other than the Library of Congress.
Contents	5. Diffraction of electromagnetic waves: 5.1. The scalar theory of diffraction; 5.2. Fresnel and Fraunhofer approximations; 5.3. Representation of the diffracted field by the angular spectrum of plane waves; 5.4. The vector theory of diffraction -- 6. Image formation in an optical system: 6.1.

Heuristic imaging theory; 6.2. The lithographic imaging system; 6.3. Imaging by a point source: coherent imaging; 6.4. Imaging by an extended source: partially coherent imaging; 6.5. Yamazoe's stacked shifted pupil formulation; 6.6. Imaging with a higher numerical aperture: radiometric correction; 6.7. 3-D point spread function; 6.8. Imaging with a higher numerical aperture: polarization effect -- 7. Aberrations in optical imaging systems: 7.1. Design of aspherical surfaces; 7.2. Connection between ray and wave aberrations; 7.3. The wave aberration for rotationally symmetric optical systems: Seidel aberrations; 7.4. General form of the aberration function for rotationally symmetric optical systems; 7.5. Digression: Gram-Schmidt orthogonalization process and orthogonal polynomials; 7.6. Orthogonal polynomials for expanding the aberration function: Zernike polynomials -- Index.

Subjects	Optics. Imaging systems. Nanolithography. Imaging systems. Nanolithography. Optics.
Notes	"SPIE Press books." Includes bibliographical references and index.
Additional formats	Online version available. http://dx.doi.org/10.1117/3.2314953
Series	SPIE Press monograph; PM284 SPIE Press monograph; PM284.

Physics: An Illustrated History of the Foundations of Science

LCCN	2016427644
Type of material	Book
Personal name	Jackson, Tom, 1972- author.
Main title	Physics: an illustrated history of the foundations of science / Tom Jackson.
Published/Produced	New York: Shelter Harbor Press, [2013]
Description	144 pages: illustrations (chiefly color); 29 cm + 1 fold-out timeline.
ISBN	9780985323066 (hbk.) 098532306X (hbk.)
LC classification	QC7.J26 2013
Summary	Presents a history of physics from the dawn of science to the present through coverage of one hundred scientific breakthroughs in the discipline, including force and inertia, hidden heat, the Doppler effect, cloud chambers, and string theory.
Contents	The dawn of science: -- Explaining nature; Thales, the father; Atoms: starting small; Four elements and more; Eureka! The Archimedes Principle; Making machinery; Seeing beams of light; The mechanics; Force and inertia; Artificial rainbows; Ockham's Razor; Adding impetus; Theory of tides; Understanding magnets; Law of refraction; Galileo: the fall guy; Applying pressure; Pendulums; Hooke's Law; Gas laws -- The scientific revolution: -- Newton's principles; Theories of light; The flying boy: conducting electricity; Temperature scale; The Leyden Jar; Hidden heat; Fire and matter; Measuring charge; Weighing a planet; Frogs legs and piles; Atomic theory; Light is a wave; Plastic and elastic; Electricity meets magnets; The thermolectric effect; Heat engines; Brownian motion -- From classical to modern physics: -- Inducing currents; The Doppler

effect; Thermodynamics: the first law; The mechanical equivalent of heat; One energy; Absolute temperature; Working at light speed; Spectroscopy: essential information; Maxwell's equations; Going from hot to cold; Electrifying gases; Boltzmann's equation; Tesla: an alternating character; Mach goes supersonic; Looking for ether; Waves through nothing; X: the unknown ray -- The subatomic age: -- Radioactivity; The first subatomic particle; Planck's constant; Long-distance radio; The Curies; Einstein's annus mirabilis; Special relativity; A positive discovery; Units of charge; Cloud chambers; Superconductors; Cosmic rays; The quantum atom; General relativity: space and time; The proton; Wave-particle duality; Exclusion principle; Bosons: force particles; An uncertain universe; Geiger's counter; Antimatter: the same but different; Atom smasher; The electron microscope; Neutrons: the final piece; Positrons: a new puzzle; Missing matter; Indoor lightning; Speeding light: Cherenkov radiation; Exotic particles; Superfluidity; Nuclear fission -- Modern physics: -- QED: quantum electrodynamics; Transistors; The big bang; Bubbles and sparks; Ivy Mike: another big bang; Masers, then lasers; Neutrino flavors; Quarks: strangeness and charm; The Standard Model; String theory; Hawking radiation; Spintronics; Dark energy; The hunt for the Higgs; Supersymmetry? -- 101 physics: the basics -- Imponderables -- The great physicists -- Bibliography and other resources -- Index -- Acknowledgments -- A timeline history of physics (back pocket) -- Measuring the universe -- Inside matter.

Subjects Physics--History.

	Physics--Popular works.
	Physics.
Form/Genre	History.
	Popular works.
Notes	Includes A timeline history of physics in back pocket.
	Includes bibliographical references (page 140) and index.
Series	Ponderables
	Ponderables (Series)

Physics on Your Feet: Berkieley Graduate Exam Questions, or, Ninety Minutes of Shame but a PhD for the Rest of Your Life!

LCCN	2014944463
Type of material	Book
Personal name	Budker, Dmitry, author.
Main title	Physics on your feet: Berkieley graduate exam questions, or, Ninety minutes of shame but a PhD for the rest of your life! / Dmitry Budker, University of California, Berkeley, USA, and Johannes Gutenberg University, Mainz, Germany, Alexander O. Sushkov, Harvard University, Cambridge, MA, USA; illustrated by Vasiliki Demas.
Edition	First edition.
Published/Produced	Oxford: Oxford University Press, 2015.
Description	xiii, 203 pages: illustrations; 25 cm
ISBN	9780199681655 (hbk.: alk. paper)
	0199681651 (hbk.: alk. paper)
	9780199681662 (pbk.: alk. paper)
	019968166X (pbk.: alk. paper)
LC classification	QC24.5.B83 2015
Portion of title	Berkieley graduate exam questions
Other title	Ninety minutes of shame but a PhD for the rest of

	your life!
Related names	Sushkov, Alexander O., author. Demas, Vasiliki, illustrator.
Summary	The authors, previously students in the Physics Department of the University of California at Berkeley, present possible questions from Berkeley's former oral examination for first year graduate students. The questions are followed by answers, all done with humor, accompanied by illustrations and cartoons.
Contents	Mechanics, heat, and general physics -- Fluids -- Gravitation, astrophysics, and cosmology -- Electromagnetism -- Optics -- Quantum, atomic, and molecular physics -- Nuclear and elementary-particle physics -- Solid-state physics -- Appendix A. Maxwell's equations and electromagnetic field boundary conditions -- Appendix B. Symbols and useful constants.
Subjects	University of California, Berkeley. Department of Physics. Physics--Popular works. Physics--Examinations, questions, etc. Physics--Humor.
Notes	Includes bibliographical references (pages 199-200) and index.

Polarized Light for Scientists and Engineers

LCCN	2012902207
Type of material	Book
Personal name	Collett, Edward, 1934-
Main title	Polarized light for scientists and engineers / Edward Collett, Beth Schaefer.
Published/Created	Long Branch, N.J.: The Polawave Group, Inc., 2012.

Description	xlv, 802 p.: ill.; 27 cm.
ISBN	9780967716701 0967716705
LC classification	QC441.C64 2012
Related names	Schaefer, Beth.
Contents	Preface -- A historical note -- First pages of papers by G.G. Stokes and E. Wolf -- Prologue -- Introduction -- The wave equation in classical optics -- The polarization ellipse -- The Poincaré sphere -- The Stokes polarization parameters -- Mueller matrices for polarizing components -- Polarization measurements -- Mueller matrices for reflection and transmission -- Muller matrices for dielectric plates -- The hybrid polarization sphere -- Applications of the hybrid polarization sphere -- Optical depolarizers abd scramblers -- Fresnel-Arago interference laws -- The Jones matrix calculus -- Polarization and coherence -- The electrodynamic foundation of polarized light -- Maxwell's equations for the electromagnetic field -- The classical radiation field -- The electromagnetic radiation of accelerating charges -- The classical Zeeman effect -- Relativistic radiation and scattering -- Stokes parameters for quantum systems -- Optical isolators amd optical circulators -- Crystal optics -- The fluorescence polarization of solutions -- The optics of metals -- Ellipsometry -- Appendix I. Vector representation of the optical field: application to optical activity -- Appendix II. Stokes vectors for various polarization states -- Appendix III. Mueller matrices for various polarization states -- Appendix IV. Jones matrices Stokes vectors for various polarization states.

Subjects	Polarization (Light) Polarization (Light)--Industrial applications. Polarisiertes Licht.

Polarized Light

LCCN	2010044039
Type of material	Book
Personal name	Goldstein, Dennis H.
Main title	Polarized light / Dennis H. Goldstein.
Edition	3rd ed.
Published/Created	Boca Raton, FL: CRC Press, 2011.
Description	xxi, 770 p.: ill.; 26 cm.
ISBN	9781439830406 (hardback)
LC classification	QC441.G65 2011
Summary	"An in-depth exploration of polarized light, this book covers production and uses, facilitating self-study without prior knowledge of Maxwell's equations. This third edition includes more than 2500 figures and equations along with chapters on polarization elements, anisotropic materials, Stokes polarimetry, Mueller matrix polarimetry, and the mathematics of the Mueller matrix. It features two new chapters, one on polarized light in nature and one on birefringence. It also presents a completely revised review of the history of polarized light and contains a new appendix on conventions used in polarized light"-- Provided by publisher.
Subjects	Polarization (Light) Polarization (Light)--Industrial applications. Technology & Engineering / Lasers & Photonics Science / Physics
Notes	Includes bibliographical references (p. 745-746) and index.

Primary Theory of Electromagnetics

LCCN	2013942979
Type of material	Book
Personal name	Eom, Hyo J., 1950-
Main title	Primary theory of electromagnetics / Hyo J. Eom.
Published/Produced	Dordrecht; New York: Springer, [2013] ©2013
Description	ix, 205 pages: illustrations; 25 cm.
ISBN	9789400771420 (alk. paper) 9400771428 (alk. paper)
LC classification	QC760.4.M37 E66 2013
Summary	"This is a textbook on electromagnetics for undergraduate students in electrical engineering, information, and communications. The book contents are very compact and brief compared to other commonly known electromagnetic books for undergraduate students and emphasizes mathematical aspects of basic electromagnetic theory. The book presents basic electromagnetic theory starting from static fields to time-varying fields. Topics are divided into static electric fields, static magnetic fields, time-varying fields, and electromagnetic waves. The goal of this textbook is to lead students away from memorization, but towards a deeper understanding of formulas that are used in electromagnetic theory. Many formulas commonly used for electromagnetic analysis are mathematically derived from a few empirical laws. Physical interpretations of formulas are de-emphasized. Each important formula is framed to indicate its significance. Primary Theory of Electromagnetics shows a clear and rigorous account of formulas in a consistent manner, thus letting students understand how

	electromagnetic formulas are related to each other"-- Cover.
Contents	Vectors -- Electrostatics -- Magnetostatics -- Faraday's Law of Induction -- Maxwell's Equations -- Uniform Plane Waves -- Transmission Lines -- Waveguides and Antennas.
Subjects	Electromagnetism--Mathematics.
	Electromagnetism--Mathematics.
Notes	Includes index.
Series	Power systems, 1612-1287
	Power systems, 1612-1287

Principles of Discrete Time Mechanics

LCCN	2013039555
Type of material	Book
Personal name	Jaroszkiewicz, George, author.
Main title	Principles of discrete time mechanics / George Jaroszkiewicz, University of Nottingham.
Published/Produced	Cambridge, United Kingdom; New York: Cambridge University Press, 2014.
	©2014
Description	xiv, 365 pages: illustrations; 26 cm.
Links	Table of contents only http://www.loc.gov/catdir/enhancements/fy1408/2013039555-t.html
	Contributor biographical information http://www.loc.gov/catdir/enhancements/fy1408/2013039555-b.html
	Publisher description http://www.loc.gov/catdir/enhancements/fy1408/2013039555-d.html
ISBN	9781107034297
	1107034299
LC classification	QC174.13.J37 2014
Contents	The physics of discreteness -- The road to calculus -- Temporal discretization -- Discrete time

dynamics architecture -- Some models -- Classical cellular automata -- The action sum -- Worked examples -- Lee's approach to discrete time mechanics -- Elliptic billiards -- The construction of system functions -- The classical discrete time oscillator -- Type 2 temporal discretization -- Discrete time quantum mechanics -- The quantized discrete time oscillator -- Path integrals -- Quantum encoding -- Discrete time classical field equations -- The discrete time Schrodinger equation -- The discrete time Klein-Gordon equation -- The discrete time Dirac equation -- Discrete time Maxwell's equations -- The discrete time Skyrme model -- Discrete time quantum field theory -- Interacting discrete time scalar fields -- Space, time and gravitation -- Causality and observation -- Concluding remarks.

Subjects Quantum theory--Mathematics.
Mechanics.

Notes Includes bibliographical references (pages 353-360) and index.

Series Cambridge monographs on mathematical physics
Cambridge monographs on mathematical physics.

Principles of Physics: For Scientists and Engineers

LCCN 2012947066

Type of material Book

Personal name Radi, Hafez A.

Main title Principles of physics: for scientists and engineers / Hafez A. Radi, John O Rasmussen.

Published/Produced Berlin; New York: Springer, [2013]
©2013

Description xviii, 1067 pages: illustrations (some color); 24 cm.

ISBN	9783642230257 (pbk.: alk. paper) 3642230253 (pbk.: alk. paper)
LC classification	QC23.2.R33 2013
Related names	Rasmussen, John O.
Contents	Part 1. Fundamental Basics -- Dimensions and Units -- Vectors -- Part 2. Mechanics -- Motion in One Dimension -- Motion in Two Dimensions -- Force and Motion -- Work, Energy, and Power -- Linear Momentum, Collisions, and Center of Mass -- Rotational Motion -- Angular Momentum -- Mechanical Properties of Matter -- Part 3. Introductory Thermodynamics -- Thermal Properties of Matter -- Heat and the First Law of Thermodynamics -- Kinetic Theory of Gases -- Part 4. Sound and Light Waves -- Oscillations and Wave Motion -- Sound Waves -- Superposition of Sound Waves -- Light Waves and Optics -- Interference, Diffraction and Polarization of Light -- Part 5. Electricity -- Electric Force -- Electric Fields -- Gauss's Law -- Electric Potential -- Capacitors and Capacitance -- Electric Circuits -- Part 6. Magnetism -- Magnetic Fields -- Sources of Magnetic Field -- Faraday's Law, Alternating Current, and Maxwell's Equations -- Inductance, Oscillating Circuits, and AC Circuits.
Subjects	Physics. Physics.
Notes	Includes index.
Series	Undergraduate lecture notes in physics, 2192-4791 Undergraduate lecture notes in physics.

Relativistic Quantum Physics: From Advanced Quantum Mechanics to Introductory Quantum Field Theory

LCCN	2011018860
Type of material	Book
Personal name	Ohlsson, Tommy, 1973-
Main title	Relativistic quantum physics: from advanced quantum mechanics to introductory quantum field theory / Tommy Ohlsson.
Published/Created	Cambridge, UK; New York: Cambridge University Press, 2011.
Description	xii, 297 p.; 26 cm.
Links	Cover image http://assets.cambridge.org/97805217/67262/cover/9780521767262.jpg Contributor biographical information http://www.loc.gov/catdir/enhancements/fy1205/2011018860-b.html Publisher description http://www.loc.gov/catdir/enhancements/fy1205/2011018860-d.html Table of contents only http://www.loc.gov/catdir/enhancements/fy1205/2011018860-t.html
ISBN	9780521767262 0521767261
LC classification	QC174.12.O35 2011
Summary	"Quantum physics and special relativity theory were two of the greatest breakthroughs in physics during the twentieth century and contributed to paradigm shifts in physics. This book combines these two discoveries to provide a complete description of the fundamentals of relativistic quantum physics, guiding the reader effortlessly from relativistic quantum mechanics to basic quantum field theory. The book gives a thorough and detailed treatment of the subject, beginning with the classification of particles, the Klein-

Gordon equation and the Dirac equation. It then moves on to the canonical quantization procedure of the Klein-Gordon, Dirac and electromagnetic fields. Classical Yang-Mills theory, the LSZ formalism, perturbation theory, elementary processes in QED are introduced, and regularization, renormalization and radiative corrections are explored. With exercises scattered through the text and problems at the end of most chapters, the book is ideal for advanced undergraduate and graduate students in theoretical physics"-- Provided by publisher.

Contents 1. Introduction to relativistic quantum mechanics; 2. The Klein-Gordon equation; 3. The Dirac equation; 4. Quantization of the non-relativistic string; 5. Introduction to relativistic quantum field theory: propagators, interactions, and all that; 6. Quantization of the Klein-Gordon field; 7. Quantization of the Dirac field; 8. Maxwell's equations and quantization of the electromagnetic field; 9. The electromagnetic Lagrangian and introduction to Yang-Mills theory; 10. Asymptotic fields and the LSZ formalism; 11. Perturbation theory; 12. Elementary processes of quantum electrodynamics; 13. Introduction to regularization, renormalization, and radiative corrections; Appendix; Index.

Subjects Quantum theory.

Notes Includes bibliographical references and index.

Six Ideas that Shaped Physics. Unit E, Electric and Magnetic Fields Are Unified

LCCN	2015048199
Type of material	Book
Personal name	Moore, Thomas A. (Thomas Andrew), author.
Main title	Six ideas that shaped physics. Unit E, Electric and magnetic fields are unified / Thomas A. Moore.
Edition	Third edition.
Published/Produced	New York, NY: McGraw-Hill Education, [2016] ©2017
Links	Publisher description http://www.loc.gov/catdir/enhancements/fy1606/2015048199-d.html
	Table of contents only http://www.loc.gov/catdir/enhancements/fy1606/2015048199-t.html
ISBN	9780077600921 (alk. paper)
	0077600924 (alk. paper)
LC classification	QC665.E4 M66 2016
Portion of title	Electric and magnetic fields are unified
Contents	Electric fields -- Charge distributions -- Electric potential -- Static equilibrium -- Current -- Dynamic equilibrium -- Analyzing circuits -- Magnetic fields -- Currents respond to magnetic fields -- Currents create magnetic fields -- The electromagnetic field -- Gauss's law -- Ampere's law -- Integral forms -- Maxwell's equations -- Faraday's law -- Induction -- Electromagnetic waves.
Subjects	Electromagnetic fields--Textbooks.
Notes	Includes index.

Solar System Astrophysics: Planetary Atmospheres and the Outer Solar System

LCCN	2013949767
Type of material	Book
Personal name	Milone, E. F., 1939- author.
Main title	Solar system astrophysics: planetary atmospheres and the outer solar system / Eugene F. Milone, William J.F. Wilson.
Edition	Second edition.
Published/Produced	New York: Springer, [2014]
Description	xviiii, 337-818 pages: illustrations (some color); 24 cm
ISBN	9781461490890 (alk. paper) 1461490898 (alk. paper)
LC classification	QB461.M552 2014
Related names	Wilson, William J. F., author.
Summary	The second edition of Solar System Astrophysics: Planetary Atmospheres and the Outer Solar System provides a timely update of our knowledge of planetary atmospheres and the bodies of the outer solar system and their analogs in other planetary systems. This volume begins with an expanded treatment of the physics, chemistry, and meteorology of the atmospheres of the Earth, Venus, and Mars, moving on to their magnetospheres and then to a full discussion of the gas and ice giants and their properties. From here, attention switches to the small bodies of the solar system, beginning with the natural satellites. Then comets, meteors, meteorites, and asteroids are discussed in order, and the volume concludes with the origin and evolution of our solar system. Finally, a fully revised section on extrasolar planetary systems puts the development of our

system in a wider and increasingly well understood galactic context. All of the material is presented within a framework of historical importance. This book and its sister volume, Solar System Astrophysics: Background Science and the Inner Solar system, are pedagogically well written, providing clearly illustrated explanations, for example, of such topics as the numerical integration of the Adams-Williamson equation, the equations of state in planetary interiors and atmospheres, Maxwells equations as applied to planetary ionospheres and magnetospheres, and the physics and chemistry of the Habitable Zone in planetary systems. Together, the volumes form a comprehensive text for any university course that aims to deal with all aspects of solar and extra-solar planetary systems. They will appeal separately to the intellectually curious who would like to know just how far our knowledge of the solar system has progressed in recent years. -- Source other than Library of Congress.

Contents Planetary Atmospheres -- Planetary Ionospheres and Magnetospheres -- The Giant Planets -- Satellite and Ring Systems -- Comets and Meteors -- Meteorites, Asteroids, and the Age and Origin of the solar System -- Extra-solar Planetary Systems.

Subjects Astrophysics.
Planets--Atmospheres.
Planets--Ionospheres.
Planets--Magnetospheres.
Comets.
Meteors.
Meteorites.

Asteroids.
Planets--Origin.
Extrasolar planets.
Planetary science.
Asteroids.
Astrophysics.
Comets.
Extrasolar planets.
Meteorites.
Meteors.
Planetology.
Planets--Atmospheres.
Planets--Ionospheres.
Planets--Magnetospheres.
Planets--Origin.
Solar system.
Solar system.

Notes "As in the first edition [2008], we maintain the two-volume bifurcation of the inner and outer regions of the solar system." --Preface to the Second Edition.
Subtitle of first volume: "Background Science and the Inner Solar System."
Includes bibliographical references and index.

Series Astronomy and astrophysics library 0941-7834
Astronomy and astrophysics library.

Special Relativity: A Heuristic Approach

LCCN 2017938702

Type of material Book

Personal name Hassani, Sadri, author.

Main title Special relativity: a heuristic approach / Sadri Hassani, University of Illinois at Urbana-Champaign, Urbana, IL, USA, Illnois State

University, Normal,IL, USA.
Published/Produced Amsterdam, Netherlands: Elsevier, 2017.
Description xix, 360 pages: illustrations; 24 cm
ISBN 9780128104118 (pbk)
0128104112 (pbk)
LC classification QC173.65.H37 2017
Contents Preface -- List of Symbols, Phrases, and Acronyms -- Note to the Reader -- 1. Qualitative Relativity -- 2. Relativity of Time and Space -- 3. Lorentz Transformation -- 4. Spacetime Geometry -- 5. Spacetime Momentum -- 6. Relativity in Four Dimensions -- 7. Relativistic Photography -- 8. Relativistic Interactions -- 9. Interstellar Travel -- 10. A Painless Introduction to Tensors -- 11. Relativistic Electrodynamics -- 12. Early Universe -- Appendix A. Maxwell's Equations -- Appendix B. Derivation of 4D Lorentz Transformation -- Appendix C. Relativistic Photography Formulas -- Bibliography -- Index.
Subjects Special relativity (Physics)
Special relativity (Physics)
Notes Includes bibliographical references and index.

Special Relativity and Classical Field Theory: The Theoretical Minimum
LCCN 2017935228
Type of material Book
Personal name Susskind, Leonard, author.
Main title Special relativity and classical field theory: the theoretical minimum / Leonard Susskind and Art Friedman.
Edition First edition.
Published/Produced New York: Basic Books, [2017]
©2017

Description	xx, 425 pages: illustrations; 22 cm.
ISBN	9780465093342 (hardback) 0465093345 (hardback) (electronic book)
LC classification	QC173.65.S87 2017
Related names	Friedman, Art, author.
Summary	"Physicist Leonard Susskind and data engineer Art Friedman are back. This time, they introduce readers to Einstein's special relativity and Maxwell's classical field theory. Using their typical brand of real math, enlightening drawings, and humor, Susskind and Friedman walk us through the complexities of waves, forces, and particles by exploring special relativity and electromagnetism. It's a must-read for both devotees of the series and any armchair physicist who wants to improve their knowledge of physics' deepest truths."-- Amazon.com.
Contents	Introduction -- The Lorentz transformation -- Velocities and 4-vectors -- Relativistic laws of motion -- Classical field theory -- Particles and fields -- Crazy units -- The Lorentz Force Law -- Fundamental principles and gauge invariance -- Maxwell's equations -- Physical consequences of Maxwell's equations -- Maxwell from Lagrange -- Fields and classical mechanics -- Appendix A. Magnetic monopoles: Lenny fools Art -- Appendix B. Review of 3-vector operators.
Subjects	Special relativity (Physics) Field theory (Physics) Science / Physics / General. Science--Relativity. Field theory (Physics) Special relativity (Physics)

Notes	"September 2017"--Title page verso. Includes index.
Series	Theoretical minimum; third volume Susskind, Leonard. Theoretical minimum; v. III.

The Art and Science of Ultrawideband Antennas

LCCN	2015451649
Type of material	Book
Personal name	Schantz, Hans.
Main title	The art and science of ultrawideband antennas / Hans Schantz.
Edition	Second edition.
Published/Produced	Boston Artech House, [2015]
Description	xxi, 563 pages: illustrations; 24 cm.
ISBN	1608079554 9781608079551
LC classification	TK7871.67.U45 S33 2015
Variant title	Ultrawideband antennas
Contents	1.1. Antennas and Elephants -- 1.2. Three Centuries of UWB Antennas -- 1.3. What Is an Antenna? -- 1.3.1. Antennas as Transducers -- 1.3.2. Antennas as Transformers -- 1.3.3. Antennas as Radiators -- 1.3.4. Antennas as Energy Converters -- 1.4. UWB Antennas -- 1.4.1. A Taxonomy of UWB Antennas -- 1.4.2. UWB Device and Systems Considerations -- 1.5. Conclusion -- Endnotes -- 2.1. Nineteenth-Century UWB Antennas -- 2.1.1. Heinrich Hertz -- 2.1.2. Narrowband in Conception, UWB in Practice -- 2.1.3. Jagadis Chandra Bose -- 2.1.4. Oliver Lodge -- 2.1.5. Guglielmo Marconi -- 2.2. Twentieth-Century UWB Antennas -- 2.2.1. Rediscovery of the Biconical Antenna -- 2.2.2. Bulbous UWB Elements -- 2.2.3. Rediscovery of the Horn

Antenna -- 2.2.4. Toward More Manufacturable Designs -- 2.2.5. Frequency-Independent Antennas -- 2.2.6. Origins of Ultrawideband Radio Technology -- 2.3. Twenty-First-Century UWB Antennas -- 2.3.1. Planar Monopole Antennas -- 2.3.2. Planar Dipole Antennas -- 2.3.3. Magnetic Antennas -- 2.3.4. Frequency-Notched UWB Antennas -- 2.3.5. Other Recent Advances -- 2.3.6. Progress in UWB Commercialization -- 2.4. UWB Antennas: What's Ahead? -- 2.5. Conclusions -- Problems -- Endnotes -- 3.1. Bandwidth -- 3.1.1. Calculating Bandwidth -- 3.1.2. Determining Antenna Bandwidth -- 3.1.3. Radiated Bandwidth -- 3.2. Dispersion -- 3.2.1. Example of a Dispersive Antenna -- 3.2.2. Example of a Nondispersive Antenna -- 3.2.3. Angular Dependence of Dispersion -- 3.3. Where Energy Goes -- 3.3.1. Antenna Pattern -- 3.3.2. Antenna Directivity, Gain, and Bandwidth -- 3.4. Pattern, Gain, and UWB Antennas -- 3.4.1. Reciprocity and UWB Antennas -- 3.4.2. Constant Gain Antennas -- 3.4.3. Constant Aperture Antennas -- 3.4.4. Other UWB Antennas -- 3.4.5. Gain and Aperture -- 3.5. Polarization -- 3.6. Antenna Matching -- 3.7. Antennas as Transducers -- Problems -- Endnotes -- 4.1. Introduction to Antenna Impedance -- 4.1.1. UWB Versus Narrowband Antenna Impedance -- 4.1.2. Controlling Antenna Impedance -- 4.2. Transmission Lines -- 4.2.1. Early Developments -- 4.2.2. Twin-Lead Transmission Line -- 4.2.3. Coaxial Transmission Line -- 4.2.4. Parallel Plane Transmission Line -- 4.2.5. Microstrip Line -- 4.3. Transition from Feed Line to Free Space -- 4.3.1.

Twin-Lead Transition -- 4.3.2. Coaxial Transitions -- 4.3.3. Planar Transmission Line Transitions -- 4.4. Impedance Transformation and Matching -- 4.4.1. The Terminated, Lossless Line -- 4.4.2. Time-Domain Reflectometry -- 4.4.3. Harmonic Signals -- 4.4.4. The Smith Chart and Matching Networks -- 4.4.5. Broadband Matching -- 4.5. Coupling Balanced and Unbalanced Lines -- 4.5.1. Chokes -- 4.5.2. Balun Transformers -- 4.5.3. Compatibility -- 4.6. Antennas as Transformers -- Problems -- Endnotes -- 5.1. Time Domain and Frequency Domain -- 5.1.1. Impulses and Sine Waves -- 5.1.2. Basic Principles of the Frequency and Time Domain -- 5.1.3. Time-Domain Signals -- 5.1.4. Time Domain Versus Frequency Domain -- 5.2. Maxwell's Equations -- 5.2.1. Generalized Coordinates and Retardation -- 5.2.2. Electromagnetic Waves -- 5.2.3. Jefimenko Form of the Biot-Savart and Coulomb Laws -- 5.2.4. Right-Hand Rule for Radiation -- 5.2.5. Time-Domain Representation of Plane Waves -- 5.3. Linear Antennas -- 5.3.1. Linear Antenna Behavior -- 5.3.2. Examples -- 5.3.3. Summary -- 5.4. Dipole Fields -- 5.4.1. Electric Dipole Fields -- 5.4.2. Magnetic Dipole Fields -- 5.4.3. Harmonic Dipole Fields -- 5.4.4. Exponentially Decaying Dipole Fields -- 5.5. Basic Antenna Physics -- 5.5.1. Antenna Differentiation -- 5.5.2. The Radiation Field Approximation -- 5.5.3. Radiation of a DC Signal? -- 5.5.4. Field Lines -- 5.6. Antennas as Radiators -- Problems -- Endnotes -- 6.1. Motivation -- 6.1.1. Models and Reality -- 6.1.2. Is Radiation "Kinky"? -- 6.1.3. The Maxwellian Perspective -- 6.2. Localization and

Flow of Electromagnetic Energy -- 6.2.1. Localizing Electromagnetic Energy -- 6.2.2. The Flow of Electromagnetic Energy -- 6.2.3. Puzzles and Paradoxes of Electromagnetic Energy Flow -- 6.2.4. Electromagnetic Velocity -- 6.2.5. Causal Surfaces -- 6.3. Electromagnetic Energy in Simple Problems -- 6.3.1. Energy in Superposition -- 6.3.2. Energy in Standing Waves -- 6.4. Dipole Field Energy -- 6.4.1. Exponentially Decaying Dipoles -- 6.4.2. Damped Harmonic Dipoles -- 6.4.3. Harmonic Dipoles -- 6.4.4. Time-Domain Excitations -- 6.5. Optimal Element Design -- 6.5.1. Fatter Is Better -- 6.5.2. Optimal Dipole Shape -- 6.5.3. Optimal Loop Shape -- 6.6. Fundamental Limits on Antenna Size -- 6.6.1. The Chu-Harrington Limit -- 6.6.2. The McLean Derivation of Small Antenna Q -- 6.6.3. Is There a Q in UWB? -- 6.6.4. Q-Based Antenna Limits in UWB Practice -- 6.6.5. Energy-Flow-Based Limits to Antenna Performance -- 6.7. Antennas as Energy Converters -- Problems -- Endnotes -- 7.1. Frequency-Independent Antennas -- 7.1.1. Basic Principles of Frequency-Independent Antennas -- 7.1.2. Spiral Antennas -- 7.1.3. Log-Periodic Antennas -- 7.1.4. Fractal and Bent Wire Antennas -- 7.2. Small-Element Electric Antennas -- 7.2.1. Conical Antennas -- 7.2.2. Planar Conical Antennas -- 7.2.3. Bulbous Antennas -- 7.2.4. Planar Bulbous Antennas -- 7.2.5. Other Planar Monopole UWB Antennas -- 7.2.6. Physical Behavior of Planar UWB Antennas -- 7.2.7. General Principles of Small-Element Design -- 7.2.8. Summary of Small-Element Electric Antennas -- 7.3. Small-Element Magnetic

Antennas -- 7.3.1. Complementarity -- 7.3.2. UWB Slot Antennas -- 7.3.3. Large Current Radiator Antennas -- 7.3.4. Monoloop Antennas -- 7.3.5. Loop Antennas -- 7.3.6. Summary of Small-Element Magnetic Antennas -- 7.4. Electrically Small Antennas -- 7.4.1. Antenna Scaling -- 7.4.2. Dielectric Loading -- 7.4.3. Conducting Enclosure Antennas -- 7.4.4. Electric-Magnetic Antenna -- 7.4.5. Summary of Electrically-Small antennas -- 7.5. Directional Electrically Small Antennas -- 7.5.1. The Beverage Loop -- 7.5.2. Ewes, Flags, and K9AYs -- 7.5.3. Multipole Synthesis -- 7.5.4. Performance of Multipole Designs -- 7.5.5. Experimental Results -- 7.6. Horn Antennas -- 7.6.1. Conical Plate Horn Antenna -- 7.6.2. Termination of Horn Antennas -- 7.6.3. Planar Horn Antennas -- 7.6.4. Other Horn Antennas -- 7.6.5. Summary of Horn Antennas -- 7.7. Reflector Antennas -- 7.7.1. Planar Reflector -- 7.7.2. Corner Reflectors -- 7.7.3. Parabolic Cylinder Reflectors -- 7.7.4. Impulse Radiating Antennas -- 7.7.5. Summary of Reflector Antennas -- 7.8. Summary -- Problems -- Endnotes -- 8.1. Antenna Spectral Control -- 8.1.1. Antenna Scaling -- 8.1.2. Antenna Filtering -- 8.1.3. Antennas and Spectral Control -- 8.2. Antenna Efficiency -- 8.2.1. Efficiency Theory -- 8.2.2. Efficiency Measurement -- 8.3. Antenna Directivity -- 8.3.1. Omni Versus Directional -- 8.3.2. Amplitude Comparison Direction Finding -- 8.3.3. Small-Aperture UWB Direction Finding -- 8.3.4. Applications -- 8.3.5. Conclusion -- 8.4. UWB Antennas in Systems -- 8.5. Summary and Conclusions -- Problems -- Endnotes.

Subjects	Ultra-wideband antennas.
Notes	Includes bibliographical references and index.
Series	Artech House antennas and propagation library Artech House antennas and propagation library.

The Mathematical Theory of Time-Harmonic Maxwell's Equations: Expansion, Integral, and Variational Methods

LCCN	2014949883
Type of material	Book
Personal name	Kirsch, Andreas.
Main title	The mathematical theory of time-harmonic Maxwell's equations: expansion, integral, and variational methods / Andreas Kirsch.
Published/Produced	New York: Springer, 2014.
Links	Table of contents only http://www.loc.gov/catdir/enhancements/fy1501/2014949883-t.html Contributor biographical information http://www.loc.gov/catdir/enhancements/fy1501/2014949883-b.html Publisher description http://www.loc.gov/catdir/enhancements/fy1501/2014949883-d.html
ISBN	9783319110851

The Mimetic Finite Difference Method for Elliptic Problems

LCCN	2013951789
Type of material	Book
Personal name	Beirão da Veiga, Lourenço.
Main title	The mimetic finite difference method for elliptic problems / Lourenço Beirão da Veiga, Konstantin Lipnikov, Gianmarco Manzini.
Published/Produced	Cham; New York: Springer, [2014] ©2014
Description	xvi, 392 pages: Illustrations; 24 cm
ISBN	9783319026626 (hbk.: alk. paper)

	3319026623 (hbk.: alk. paper)
LC classification	QA374.B325 2014
Related names	Lipnikov, Konstantin. Manzini, Gianmarco.
Abstract	This book describes the theoretical and computational aspects of the mimetic finite difference method for a wide class of multidimensional elliptic problems, which includes diffusion, advection-diffusion, Stokes, elasticity, magnetostatics and plate bending problems. The modern mimetic discretization technology developed in part by the Authors allows one to solve these equations on unstructured polygonal, polyhedral and generalized polyhedral meshes. The book provides a practical guide for those scientists and engineers that are interested in the computational properties of the mimetic finite difference method such as the accuracy, stability, robustness, and efficiency. Many examples are provided to help the reader to understand and implement this method. This monograph also provides the essential background material and describes basic mathematical tools required to develop further the mimetic discretization technology and to extend it to various applications. -- Source other than Library of Congress.
Contents	1. Model elliptic problems -- 2. Foundations of mimetic finite difference method -- 3. Mimetic inner products and reconstruction operators -- 4. Mimetic discretization of bilinear forms -- 5. The diffusion problem in mixed form -- 6. The diffusion problem in primal form -- 7. Maxwell's equations -- 8. The Stokes problem -- 9. Elasticity

	and plates -- 10. Other linear and nonlinear mimetic schemes -- 11. Analysis of parameters and maximum principles -- 12. Diffusion problem on generalized polyhedral meshes.
Subjects	Differential equations, Partial. Differential equations, Partial.
Notes	Includes bibliographical references (pages 371-389) and index.
Series	MS&A, 2037-5255; 11 MS&A (Series); 11.

The Physics of Reality: Space, Time, Matter, Cosmos: Proceedings of the 8th Symposium Honoring Mathematical Physicist Jean-Pierre Vigier, Covent Garden, London, UK, 15-18 August 2012

LCCN	2014395431
Type of material	Book
Meeting name	Symposium Honoring Mathematical Physicist Jean-Pierre Vigier (8th: 2012: London, England)
Main title	The physics of reality: space, time, matter, cosmos: proceedings of the 8th symposium honoring mathematical physicist Jean-Pierre Vigier, Covent Garden, London, UK, 15-18 August 2012 / editors, Richard L. Amoroso (Noetic Advanced Studies Institute, USA), Louis H. Kauffman (University of Illinois at Chicago, USA), Peter Rowlands (University of Liverpool, UK).
Published/Produced	[Hackensack], New Jersey: World Scientific, [2013]
Description	xxvi, 526 pages illustrations; 29 cm
ISBN	9789814504775 (hbk.) 9814504777 (hbk.)
LC classification	QB980.S96 2012
Related names	Amoroso, Richard L., editor. Kauffman, Louis H., 1945- editor.

Rowlands, Peter, editor.

Contents Section I. Foundational physics. 1. Laws of form, Majorana fermions, and discrete physics / Louis H. Kauffman -- 2. The tie that binds: a fundamental unit of 'change' in space and time / James E. Beichler -- 3. Space and antispace / Peter Rowlands -- 4. Zero-totality in action-reaction space: a generalization of Newton's third law? / Sabah E. Karam -- 5. A computational unification of scientific law: spelling out a universal semantics for physical reality / Peter J. Marcer and Peter Rowlands -- 6. Eliminating the gravitational constant, G, and improving SI units, using two new Planck-unit frameworks with all parameters as ratios of only H, C, and a / Michael Lawrence -- 7. A programmable cellular-automata polarized Dirac vacuum / Drahcir S. Osoroma -- 8. Exploring novel cyclic extensions of Hamilton's dual-quaternion algebra / Richard L. Amoroso, Peter Rowlands & Louis H. Kauffman -- Section II. Special and general relativity. 9. Explicit and implicit uncertainties and the uncertainty principle in the special theory of relativity / Oleg V. Matvejev & Vadim N. Matveev -- 10. Simulations of relativistic effects, relativistic time, and the constancy of light velocity / Vadim N. Matveev & Oleg V. Matvejev -- 11. On the maximum speed of matter / Dionysios G. Raftopoulos -- 12. Spacetime and quantum propagation from digital clocks / Garnet N. Ord -- 13. General relativity theory -- well proven and also incomplete? / Jürgen Brandes -- 14. Rectification of general relativity, experimental verifications, and errors of the Wheeler School / C.Y. Lo -- 15. Relativity based on physical processes rather than space-time / Albrecht Giese -- 16. On the fundamental nature of the electromagnetic interaction / Al F. Kracklauer -- 17. Unified geometrodynamics: a complementarity of Newton's and Einstein's gravity / Richard L. Amoroso -- 18. Review of the relationship between Galilean relativity and the velocity of light / Shukri Klinaku -- 19. Review of the derivation of the Lorentz transformation / Shukri Klinaku & Valbona Berisha -- 20. Marriage of electromagnetism and gravity in an extended space model and astrophysical phenomena / V.A. Andreev & D.Yu. Tsipenyuk -- 21. Both the twin paradox and GPS data show the need for additional physics / J.N. Percival -- 22. An alternative to Dirac's model, confining change due to the bound electron, not in the field, but within the electron itself / Tolga Yarman... [et al.]. Section III. Thermodynamics, fields, and gravity. 23. Thoughts on Landauer's principle and its experimental verification / David Sands & Jeremy Dunning-Davies -- 24. Information, entropy, and the classical ideal gas / David Sands & Jeremy Dunning-Davies -- 25. Clausius' concepts of 'aequivalenzwerth' and entropy: a critical appraisal / David Sands & Jeremy Dunning-Davies -- 26. The conformal steady-state free precession: a Kepplerian approach to automorphic scattering theory of orbiton/spinon dynamics / Walter J. Schempp -- 27. An introduction to Boscovichian unified field theory / Roger James Anderton -- 28. A scintilla of unified field mechanics revealed by a conceptual integration of new fundamental elements associated with wavepacket dispersion / Richard L. Amoroso & Jeremy Dunning-Davies -- 29. Fields in action?

From the inside looking out (musings of an idiosyncratic experimentalist!) / John Dainton -- Section IV. Quantum mechanics. 30. Evidencing 'tight bound states' in the hydrogen atom: empirical manipulation of large-scale XD in violation of QED / Richard L. Amoroso & Jean-Pierre Vigier -- 31. Quantum causality / Sergey M. Korotaev & Evgeniy O. Kiktenko -- 32. Relativistic entanglement / John E. Carroll & Adrian H. Quarterman -- 33. Relativistic entanglement from Maxwell's classical equations / john e. carroll & adrian h. quarterman -- 34. Universal quantum computing: third gen prototyping utilizing relativistic 'trivector' R-qubit modeling surmounting uncertainty / Richard L. Amoroso, Louis H. Kauffman & Salvatore Giandinoto -- 35. Deterministic impulsive vacuum foundations for quantum-mechanical wavefunctions / John S. Valentine -- 36. Emergence of a new quantum mechanics by multivalued logic / Claude Gaudeau De Gerlicz... [et al.] -- 37. Dirac sea and its evolution / Boris Volfson -- 38. Justifying the vacuum as an electron-positron aggregation and experimental falsification / R. Guy Grantham & Ian G. Montgomery.Section V. Cosmology and consciousness. 39. Some thoughts on redshift and modern cosmology / Jeremy Dunning-Davies & Richard L. Amoroso -- 40. Examining the existence of the multiverse / James J. Hurtak, D.E. Hurtak & Elizabeth A. Rauscher -- 41. Universal scaling laws in quantum theory and cosmology / Elizabeth A. Rauscher, James J. Hurtak & D.E. Hurtak -- 42. Implementing Maxwell's aether illuminates the physics of gravitation: the gravity-electric (G-E) field, evident at every scale, from the ionosphere to spiral galaxies and a neutron-star extreme / Miles F. Osmaston -- 43. Continuum theory (CT): its particle-tied aether yields a continuous auto-creation, non-expanding cosmology and new light on galaxy evolution and clusters / Miles F. Osmaston -- 44. A transluminal-energy quantum model of the cosmic quantum / Richard Gauthier -- 45. Transluminal-energy quantum models of the photon and the electron / Richard Gauthier -- 46. Thoughts occasioned by the 'Big Bang' model of the universe / Jeremy Dunning-Davies -- 47. The theological basis of Big Bang cosmology and the failure of general relativity / Stephen J. Crothers -- 48. The neurogenetic correlates of consciousness / John K. Grandy -- 49. Are life, consciousness, and intelligence cosmic phenomena? / Ludwik Kostro -- 50. A holoinformational model of the physical observer / Francisco Di Biase -- 51. Time? / Richard L. Amoroso -- 52. 'Shut the front door!': obviating the challenge of large-scale extra dimensions and psychophysical bridging / Richard L. Amoroso.

Subjects	Vigier, Jean-Pierre, 1920-2004.
	Cosmology--Congresses.
	Astrophysics--Congresses.
	Mathematical physics--Congresses.
Notes	Includes bibliographical references and index.

The Power and Beauty of Electromagnetic Fields

LCCN	2011008250
Type of material	Book
Personal name	Morgenthaler, Frederic R.
Main title	The power and beauty of electromagnetic fields /

	F.R. Morgenthaler.
Published/Created	Hoboken, N.J.: Wiley, c2011.
Description	xxxiii, 639 p.: ill.; 26 cm. + 1 DVD-ROM (4 3/4 in.)
ISBN	9781118057575 (hardback) 1118057570 (hardback)
LC classification	QC665.E4 M67 2011
Summary	"In this text, the author develops alternate representations of electromagnetic power and energy that differ form the familiar Maxwell-Poynting theorem values (S and W) - yet are fully equivalent. The particular choice focuses on features highly-localized power and energy components and emphasizes the circuit rather than the wave nature of theses quantities. Moreover, unlike the Poynting vector, this exact representation merges smoothly with well-known quasistatic approximations that have long been used to calculate power flows in both lumped and distributed circuits operating at low-frequencies"--Provided by publisher.
Contents	Pt. I. Basic electromagnetic theory -- 1. Maxwell's equations -- 2. Quasistatic approximations -- 3. Electromagnetic power, energy, stress and momentum -- 4. Electromagnetic waves in free-space -- 5. Electromagnetic waves in linear materials -- 6. Electromagnetic theorems and principles -- Pt. II. Four-dimensional electromagnetism -- 7. Four-dimensional vectors and tensors -- 8. Energy-momentum tensors -- 9. Dielectric and magnetic materials -- 10. Amperian, Minkowski, and Chu formulations -- Pt. III. Electromagnetic examples -- 11. Static and quasistatic fields -- 12. Uniformly moving electric

	charges -- 13. Accelerating charges -- 14. Uniform surface current -- 15. Uniform line currents -- 16. Plane waves -- 17. Waves incident at a material interface -- 18. TEM transmission lines -- 19. Rectangular waveguide modes -- 20. Circular waveguide modes -- 21. Dielectric waveguides -- 22. Antennas and diffraction -- 23. Waves and resonances in ferrites -- 24. Equivalent circuits -- 25. Practice problems -- Pt. IV. Backmatter -- Appendices A-H.
Subjects	Electromagnetic fields.
Notes	Includes bibliographical references and index.
Series	The IEEE Press series on electromagnetic wave theory IEEE Press series on electromagnetic wave theory.

The Quantum Puzzle: Critique of Quantum Theory and Electrodynamics

LCCN	2017275091
Type of material	Book
Personal name	Clarke, Barry R., author.
Main title	The quantum puzzle: critique of quantum theory and electrodynamics / Barry R Clarke, Brunel University, UK.
Published/Produced	New Jersey: World Scientific, [2017] ©2017
Description	xiii, 386 pages: illustrations; 24 cm.
ISBN	9789814696968 (hbk) 981469696X (hbk)
LC classification	QC174.12.C52 2017
Summary	"In 1861, James Clerk-Maxwell published Part II of his four-part series 'On physical lines of force'. In it, he attempted to construct a vortex model of the magnetic field but after much effort neither he,

nor other late nineteenth century physicists who followed him, managed to produce a workable theory. What survived from these attempts were Maxwell's four equations of electrodynamics together with the Lorentz force law, formulae that made no attempt to describe an underlying reality but stood only as a mathematical description of the observed phenomena. When the quantum of action was introduced by Planck in 1900 the difficulties that had faced Maxwell's generation were still unresolved. Since then theories of increasing mathematical complexity have been constructed to attempt to bring the totality of phenomena into order. This work examines the problems that had been abandoned long before quantum mechanics was formulated in 1925 and argues that these issues need to be revisited before any confidence can be placed in a quantum theory of the electromagnetic field."-- Back cover.

Subjects Quantum theory.
Quantum electrodynamics.
Electromagnetic interactions.
Electrodynamics.

Notes Includes bibliographical references (pages 363-379) and index.

The Universe in Zero Words: The Story of Mathematics as Told through Equations

LCCN 2011936364

Type of material Book

Personal name Mackenzie, Dana.

Main title The universe in zero words: the story of mathematics as told through equations / Dana Mackenzie.

Published/Created	Princeton, N.J.: Princeton University Press, 2012.
Description	224 p.: ill. (some col.); 25 cm.
Links	Contributor biographical information http://www.loc.gov/catdir/enhancements/fy1214/2011936364-b.html Publisher description http://www.loc.gov/catdir/enhancements/fy1214/2011936364-d.html Table of contents only http://www.loc.gov/catdir/enhancements/fy1317/2011936364-t.html
ISBN	9780691152820 0691152829
LC classification	QA211.M16 2012
Portion of title	Story of mathematics as told through equations
Contents	Pt. 1. Equations of antiquity -- 1. Why we believe in arithmatic: the world's simplest equation -- 2. Resisting a new concept: the discovery of zero -- 3. The square of the hypotenuse: the Pythagorean theorem -- 4. The circle game: the discovery of [pi] -- From Zeno's paradoxes to the idea of infinity -- 6. A matter of leverage: laws of levers -- Pt. 2. Equations in the age of exploration -- 7. the stammerer's secret: Cardano's formula -- 8. Order in the heavens: Kepler's laws of planetary motion -- 9. Writing for eternity: Fermat's last theorem -- 10. An unexplored continent: the fundamental theorem of calculus -- 11. Of apples, legends... and comets: Newton's laws -- 12. The great explorer: Euler's therems -- Pt. 3 Equations in a promethean age --13. The new algebra: Hamilton and quaternions -- 14. Two shooting stars: group theory -- 15. The geometry of whales and ants: non-Euclidean geometry -- 16. In primes we trust: the prime number theorem -- 17. The idea of spectra: Fourier series -- 18. A god's-

eye view of light: Maxwell's equations -- Pt. 4. Equations in our own time -- 19. The photoelectric effect: quanta and relativity -- 20. From a bad cigar to Westminster Abbey: Dirac's formula -- 21. The empire-builder: the Chern-Gauss-Bonnet equation -- 22. A little bit infinite: the Continuum Hypothesis -- 23. Theories of chaos: Lorenz equations -- 24. Taming the tiger: the Black-Scholes equation.

Subjects	Equations--Popular works. Mathematics--History--Popular works.
Notes	Includes bibliographical references (p. 219-221) and index.

Time-Domain Finite Element Methods for Maxwell's Equations in Metamaterials

LCCN	2012953285
Type of material	Book
Personal name	Li, Jichun.
Main title	Time-domain finite element methods for Maxwell's equations in metamaterials / Jichun Li, Yunqing Huang.
Published/Created	Heidelberg: Springer, [2013]
Description	xii, 302 pages: illustrations; 24 cm.
ISBN	9783642337888 (hbk.: alk. paper) 3642337880 (hbk.: alk. paper)
LC classification	MLCM 2018/42636 (Q)
Related names	Huang, Yunqing, 1962- author.
Subjects	Maxwell equations. Metamaterials--Mathematics. Electromagnetism--Mathematics. Finite element method.
Notes	Includes bibliographical references and index.
Series	Springer series in computational mathematics; 43

Understanding Geometric Algebra for Electromagnetic Theory

LCCN	2011005744
Type of material	Book
Personal name	Arthur, John W., 1949-
Main title	Understanding geometric algebra for electromagnetic theory / John W. Arthur.
Published/Created	Hoboken, N.J.: Wiley-IEEE Press, c2011.
Description	xvi, 301 p.: ill.; 25 cm.
Links	Table of contents only http://www.loc.gov/catdir/enhancements/fy1108/2011005744-t.html Publisher description http://www.loc.gov/catdir/enhancements/fy1108/2011005744-d.html Contributor biographical information http://www.loc.gov/catdir/enhancements/fy1114/2011005744-b.html
ISBN	9780470941638
LC classification	QC670.A76 2011
Summary	"This book aims to disseminate geometric algebra as a straightforward mathematical tool set for working with and understanding classical electromagnetic theory. It's target readership is anyone who has some knowledge of electromagnetic theory, predominantly ordinary scientists and engineers who use it in the course of their work, or postgraduate students and senior undergraduates who are seeking to broaden their knowledge and increase their understanding of the subject. It is assumed that the reader is not a mathematical specialist and is neither familiar with geometric algebra or its application to electromagnetic theory. The modern approach, geometric algebra, is the mathematical tool set we should all have started out with and once the reader has a grasp of the subject, he or she cannot

fail to realize that traditional vector analysis is really awkward and even misleading by comparison"-- Provided by publisher.
"This book covers all of the information needed to design LEDs into end-products. It is a practical guide, primarily explaning how things are done by practicing engineers. Equations are used only for practical calculations, and are kept to the level of high-school algebra. There are numerous drawings and schematics showing how things such as measurements are actually made, and showing curcuits that actually work. There are practical notes and examples embedded in the text that give pointers and how-to guides on many of the book's topics"-- Provided by publisher.

Contents Preface. -- Reading Guide. -- 1. Introduction. -- 2. A Quick Tour of Geometric Algebra. -- 2.1 The Basic Rules Geometric Algebra. -- 2.2 3D Geometric Algebra. -- 2.3 Developing the Rules. -- 2.4 Comparison with Traditional 3D Tools. -- 2.5 New Possibilities. -- 2.6 Exercises. -- 3. Applying the Abstraction. -- 3.1 Space and Time. -- 3.2 Electromagnetics. -- 3.3 The Vector Derivative. -- 3.4 The Integral Equations. -- 3.5 The Role of the Dual. -- 3.6 Exercises. -- 4. Generalisation. -- 4.1 Homogeneous and Inhomogeneous Multivectors. -- 4.2 Blades. -- 4.3 Reversal. -- Understanding Geometric Algebra for Electromagnetic Theory. -- 4.4 Maximum Grade. -- 4.5 Inner and Outer Products Involving a Multivector. -- 4.6 Inner and Outer Products between Higher Grades. -- 4.7 Summary so Far. -- 4.8 Exercises. -- 5. (3+1)D Electromagnetics. -- 5.1 The Lorentz Force. -- 5.2 Maxwell's Equations in Free Space. -- 5.3

Simplified Equations. -- 5.4 The Connexion between the Electric and Magnetic Fields. -- 5.5 Plane Electromagnetic Waves. -- 5.6 Charge Conservation. -- 5.7 Multivector Potential. -- 5.8 Energy and Momentum. -- 5.9 Maxwell's Equations on Polarisable Media. -- 5.10 Exercises. -- 6. Review of (3+1)D. -- 7. Introducing Spacetime. -- 7.1 Background and Key Concepts. -- 7.2 Time as a Vector. -- 7.3 The Spacetime Basis Elements. -- 7.4 Basic Operations. -- 7.5 Velocity. -- 7.6 Different Basis Vectors and Frames. -- 7.7 Events and Hstories. -- Understanding Geometric Algebra for Electromagnetic Theory. -- 7.8 The Spacetime Form of. -- 7.9 Working with Vector Differentiation. -- 7.10 Working without Basis Vectors. -- 7.11 Classification of Spacetime Vectors and Bivectors. -- 7.12 Exercises. -- 8. Relating Spacetime to (3+1)D. -- 8.1 The Correspondence between the Elements. -- 8.2 Translations in General. -- 8.3 Introduction to Spacetime Splits. -- 8.4 Some Important Spacetime Splits. -- 8.5 What Next? -- 8.6 Exercises. -- 9. Change of Basis Vectors. -- 9.1 Linear transformations. -- Understanding Geometric Algebra for Electromagnetic Theory. -- 9.2 Relationship to Geometric Algebras. -- 9.3 Implementing Spatial Rotations and the Lorentz Transformation. -- 9.4 Lorentz Transformation of the Basis Vectors. -- 9.5 Lorentz Transformation of the Basis Bivectors. -- 9.6 Transformation of the Unit Scalar and Pseudoscalar. -- 9.7 Reverse Lorentz Transformation. -- 9.8 The Lorentz Transformation with Vectors in Component Form. -- 9.9 Dilations. -- 9.10 Exercises. -- 10. Further

Spacetime Concepts. -- 10.1 Review of Frames and Time Vectors. -- 10.2 Frames in General. -- 10.3 Maps and Grids. -- 10.4 Proper Time. -- 10.5 Proper Velocity. -- 10.6 Relative Vectors and Paravectors. -- 10.7 Frame Dependent v. Frame Independent Scalars. -- 10.8 Change of Basis for any Object in Component Form. -- 10.9 Velocity as Seen in Different Frames. -- 10.10 Frame Free Form of the Lorentz Transformation. -- 10.11 Exercises. -- Understanding Geometric Algebra for Electromagnetic Theory. -- 11. Application of Spacetime Geometric Algebra to Basic Electromagnetics. -- 11.1 The Spacetime Approach to Electrodynamics. -- 11.2 The Vector Potential and some Spacetime Splits. -- 11.3 Maxwell's Equations in Spacetime Form. -- 11.4 Charge Conservation and the Wave Equation. -- 11.5 Plane Electromagnetic Waves. -- 11.6 Transformation of the Electromagnetic Field. -- 11.7 Lorentz Force. -- 11.8 The Electromagnetic Field of a Moving Point Charge. -- 11.9 Exercises. -- 12. The Electromagnetic Field of a Point Charge Undergoing Acceleration. -- 12.1 Working with Null Vectors. -- 12.2 Finding F for a Moving Point Charge. -- 12.3 Frad in the Charge's Rest Frame. -- 12.4 Frad in the Observer's Rest Frame. -- 12. 5 Exercises. -- 13. Conclusion. -- 14. Appendices. -- 14.1 Glossary. -- 14.2 Axial v True Vectors. -- Understanding Geometric Algebra for Electromagnetic Theory. -- 14.3 Complex Numbers and the 2D Geometric Algebra. -- 14.4 The Structure of Vector Spaces and Geometric Algebras. -- 14.5 Quaternions Compared. -- 14.6 Evaluation of an Integral in Equation (5.14). --

	14.7 Formal Derivation of the Spacetime Vector Derivative. -- 15. Table and Figure Captions. -- 16. Further Reading on Geometric Algebra. -- 17. References. -- 18. Tables and Figures.
Subjects	Electromagnetic theory--Mathematics. Geometry, Algebraic. Science / Electromagnetism
Notes	Includes bibliographical references (p. 287-289) and index.
Series	IEEE Press series on electromagnetic wave theory; 38

INDEX

D

E

F

G

H

I

K

L

M

N

O

P

Q

R

S

T

U

V

W

Y